MANUAL FOR LABORATORY WORK IN
MAMMALIAN PHYSIOLOGY

MANUAL FOR LABORATORY WORK IN
MAMMALIAN PHYSIOLOGY

Third Edition, revised

by

FRED E. D'AMOUR, FRANK R. BLOOD, and DON A. BELDEN, JR.

THE UNIVERSITY OF CHICAGO PRESS

CHICAGO AND LONDON

The University of Chicago Press, Chicago 60637
The University of Chicago Press, Ltd., London

Copyright 1948, 1954, and 1965 by The University of Chicago. All rights reserved. Published 1948. Revised Edition. published 1954. Third Edition (Revised) published 1965. Third Impression 1974. Printed in the United States of America
International Standard Book Number: 0-226-13563-2
Library of Congress Catalog Card Number: 65-17285

No illustration or any part of the text may be reproduced without permission of The University of Chicago Press.

PREFACE

In many universities and colleges, especially liberal arts colleges, laboratory work in mammalian physiology is handicapped by the difficulty of obtaining laboratory animals. Although the rat is widely used in many fields of biologic research, its suitability for routine laboratory work in college physiology has not heretofore been formally demonstrated. This manual, which includes a group of 65 experiments in which the rat is used exclusively, is presented in the belief that, by thus demonstrating the usefulness of this very convenient animal for the purpose, the problem of laboratory-animal supply will be greatly simplified.

The experiments are arranged in two parts. Part I includes the "core" group (34 experiments) required of all students and which, on the basis of a two- or three-hour laboratory twice weekly, will provide for a half year's work. For more advanced students or those wishing to go further in the field, Part II, called the "special" group, including 31 experiments, is given. Both groups together provide ample material for a full year's work. The distribution of experiments among the organ systems is fairly consistent with the emphasis usually placed upon them. No experiments on the human subject or on non-mammalian forms have been included. The instructor will automatically require the performance of a number of conventional experiments on the human subject, and the rat substitutes adequately for the type of experiment (heart action and muscle-nerve physiology) for which frogs and turtles are usually employed. Although most of the experiments are "acute," i.e., are terminated at the end of the laboratory period, several are "chronic," the animal being kept alive and observed over a period of time; these may therefore have to be done out of their regular sequence. They include all experiments on the endocrine system and experiment 23 in Part II under the nervous system. It may also be noted that, in Part II, several experiments are included, such as hypophysectomy, renal hypertension, spinal reflexes, etc., whose value is primarily that of introducing the student to research technics. In fact, we believe one of the chief values derived from the course is the research interest and ability acquired by the student.

This manual is intended to be primarily a description of technics and procedures. Verbal descriptions of apparatus and of step-by-step procedures are clarified throughout with illustrations; anatomic drawings, which are frequently inadequate, are replaced by photographs. For illustrative clarity, somewhat longer incisions and more radical dissections have been made than are necessary in performing the experiment. The explanation and interpretation of results are left to the student and his instructor, to be elucidated, usually, by a series of questions and answers. A blank page will be found at the end of each experiment for this purpose and for the student's report. No formal outline for the report is given, as each instructor will, no doubt, have his favorite form. All experiments have been performed, over the years, by the authors' students; the records and results included have been obtained by them and therefore indicate what may reasonably be expected.

In addition to a regular text, several other books may be found helpful. These are: *The Anatomy of the Rat*, by Greene (Philadelphia: American Philosophical Society), and *The Rat in Laboratory Investigation*, by Griffith and Farris (Philadelphia: J. B. Lippincott Co.).

Most of the apparatus used is that found in the usual physiology laboratory; the rat's size makes only a very few special items necessary. A binocular magnifier (jeweler's loupe) is useful, and small, fine-pointed scissors and forceps are necessary. We have found ink-writing instruments more convenient than the conventional sooted drum for kymograph records. All items of equipment, both individually and in their final setup, are fully described and illustrated under the appropriate experiment, and it will be seen how readily adaptable standard surgical procedures, methods of recording, etc., are to the use of the rat. Items of equipment especially designed for use with the rat may be purchased from Phipps and Bird, Incorporated, of Richmond, Virginia.

In a number of places a particular commercial product (drug, hormone, or item of equipment) is mentioned by name. This is not to be understood as a specific recommendation of that product to the exclusion of others, but is given, first, because the product has been found to be satisfactory and, second, for the convenience of the instructor who is unfamiliar with the products available in that particular field.

FRED E. D'AMOUR   UNIVERSITY OF DENVER
FRANK R. BLOOD   VANDERBILT UNIVERSITY
DON A. BELDEN, JR.   UNIVERSITY OF DENVER

INTRODUCTION

AIMS AND OBJECTIVES

The chief reason for doing laboratory work in any science is to become personally identified with its problems by having personally engaged in attacking and solving them. A man working with his hands and thinking about what he is doing acquires thereby a personal experience which is far more effective, in granting true understanding and in its lasting imprint upon the memory, than any textbook learning can ever be. As Harvey says: "Diligent observation is therefore requisite in every science, and the senses are to be frequently appealed to. We are, I say, to strive after personal experience, not to rely on the experience of others; without which indeed no man can properly become a student of any branch of natural science."

Frequently, however, this ideal objective of laboratory work is not achieved. Either the work is dull, its details become routine and mechanical, and, although the hands are at work, the mind is idle; or, because the technic is difficult, the attention is so completely focused on problems of apparatus and procedures that the true meaning of the experiment is lost.

You will not be bored by your laboratory work in physiology. Whatever else may be said for it—that it is difficult, demanding, even dirty—it is certainly not dull. Too many things can happen—and frequently do. As to the second point, however, a word of warning is in order. This is not a "snap" course; some of the experiments are by no means easy. If you are the right type of student you will accept the difficulties as a challenge and strive diligently to overcome them. This is as it should be, but there is danger that in the struggle with technical difficulties you may lose sight of the purpose of the experiment. To obtain a perfect record is worth while, but the aim of the course is not to gain a mastery of technics but a knowledge of physiology. Therefore, you should look upon the laboratory work not as an end in itself but as a means toward an end, as an opportunity given you actually to observe the phenomena described in text and lecture; to test the truth of statements made therein; to demonstrate how varied and complex are the factors involved in producing a certain result—in short, actually to see and study, in all its complexities, the living, functioning animal body.

STUDENT RESPONSIBILITIES

Since physiology concerns the functioning of living organisms, the laboratory material to be studied must, of necessity, be alive. The right to invade the living animal body is not, however, one to be lightly granted; it can be granted only if certain responsibilities are understood and accepted. These responsibilities are, first, a most scrupulous care that the animal suffer no pain and, second, a sincere regard for the health and well-being of the animal colony.

In undertaking this course you will be dealing, perhaps for the first time, with living laboratory subjects, and it is important that you realize the existence of this moral prerequisite. The rat is a sentient being, it can feel pain. If you were to undergo a surgical operation, you would demand that an anesthetic be used; your rat is entitled to the same consideration. Details of administration and response are described later, but it is here emphasized that during the performance of an experiment you must at all times be certain that no suffering is being inflicted.

As for the health and well-being of the colony, you should assume much of the responsibility. This is not mere janitor work; in this course rats are the most important item, and you should therefore be interested in learning all you can about them. You should take pride in the appearance of the colony, in having sleek, well-fed animals in clean, sanitary quarters. Familiarity with their habits and care for their well-being will pay dividends, in a minor way, in a reduced incidence of rat bites and, in a major way, in an increased incidence of successful results in the laboratory. Apropos of rat bites, animals which have been handled frequently become very tame and almost never bite. The bite bleeds freely but is not particularly painful. Tincture of iodine may be applied, but we have never heard of a rat bite becoming infected.

STUDENT ASSIGNMENTS

In most experiments a group of three or four students work together, and success frequently depends on a great degree of co-operation. This dependence of each on the others provides salutary training in teamwork; some beautiful friendships have resulted from this necessarily close contact.

The students' assignments are made in advance, being rotated from one experiment to the next. They are as follows:

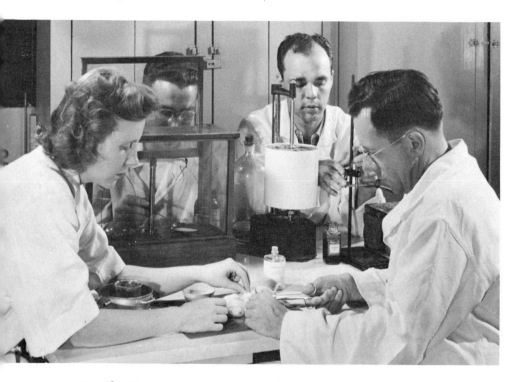

1. The Surgeon

As surgeon you perform the actual operation. It is obvious that you must know the anatomy of the region in which you will work; since the assignment is made in advance, you should make use of a previous carcass to practice the dissection. Reference to the anatomic plates in Greene's *Anatomy of the Rat* will be found invaluable. The surgeon is in complete charge of the experiment; other members accept orders from him. Since the responsibility for success is yours, the record, good or bad, becomes your property. Finally, you are responsible for seeing that the animal is certainly dead before being discarded; routinely, the thorax of every animal should be opened before it is thrown out.

2. The Assistant Surgeon and Anesthetist

If a barbiturate, such as Nembutal, is used as the basal anesthetic, you must arrive and administer it half an hour before the other members of the group appear. Thus promptness in starting the experiment depends largely on you. You must also prepare all solutions in advance and see to it that warm saline solution is ready. During the operation you will administer the ether and assist the surgeon. Even while assisting with the surgery, you are responsible for the anesthesia.

3. The Recorder

Having been assigned this position in advance, you should have familiarized yourself with the recording, stimulating, or other apparatus to be used and are responsible for having it set up and in perfect working order. If a tambour is to be used, it must be tested for leaks in ample time to replace the membrane if necessary. You must be sure that the tubes used in cannulating will fit the connections to the apparatus and that the position of the apparatus in relation to the rat will be convenient for working. The recorder can do more to expedite the performance of an experiment than any other member, merely by planning in advance.

4. General Assistant

In this position you will run errands, help anyone who needs help, keep solutions warm, and, at the conclusion of the experiment, be responsible for cleaning up. The laboratory must, of course, be left spotless, instruments washed and neatly put away, the carcass disposed of, etc. A rule that on any occasion when this has not been properly done a failure for the entire group will be recorded will insure the attention of the other members to your performance. The rule should never have to be invoked.

THE RAT COLONY

In some localities rats may be purchased from local dealers; it is much better, however, to maintain a colony on the campus for laboratory use. On the assumption that a colony is being organized for the first time, the following brief directions are given; more detailed instructions may be found in *The Rat in Laboratory Investigation* by Griffith and Farris.

It is advisable to start with good stock (rats from the Wistar Institute, Philadelphia, are recommended) and to select the best of the progeny for breeders. If work is to start in the fall, the colony should be organized in the spring. At six months of age rats will have reached nearly their full size. Litters run from six to eight, so one can calculate, on the basis of the size of the expected class, how many to start with.

The colony should be housed in a room which can be kept warm; a temperature range of from 70° to 80° F. is satisfactory. The room need not be large, two large cages for adults and a dozen or so small ones for mothers and babies are sufficient for ordinary purposes. Purina chows or a similar feed, supplemented twice a week with lettuce and carrots, forms a satisfactory diet. Water dishes should be washed daily. Wood shavings are satisfactory for bedding, and the cages should be cleaned weekly. Spraying the cages lightly with a 5 per cent D.D.T. solution once a month and scouring with soap and hot water every three months will control insects and prevent odors.

Males and females may occupy the same cage, the females being removed when obviously pregnant. Or, if controlled breeding is desired, the sexes may be kept separated and the females observed for estrus. Estrus recurs every 4–5 days and is marked by reddening and swelling of the vaginal orifice. In case of doubt, vaginal smears may be taken as described in Part I, Experiment 26. Females in estrus, placed with males overnight, will almost invariably breed. Shortly before parturition (the period of gestation is 21–22 days) they should be put into individual cages and supplied with excelsior for nest-building. The babies may be weaned after three weeks and the mothers be given a rest period of a few weeks before being mated again.

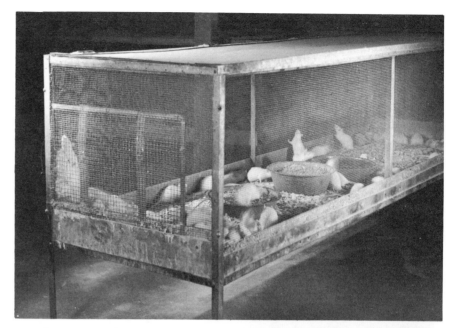

Colony cage. This cage will accommodate from 75 to 100 rats. Dimensions: 96 × 30 × 24 inches.

"Maternity" cage. This cage may be used for small groups of rats. Dimensions: 12 × 15 × 10 inches.

PART I CORE EXPERIMENTS

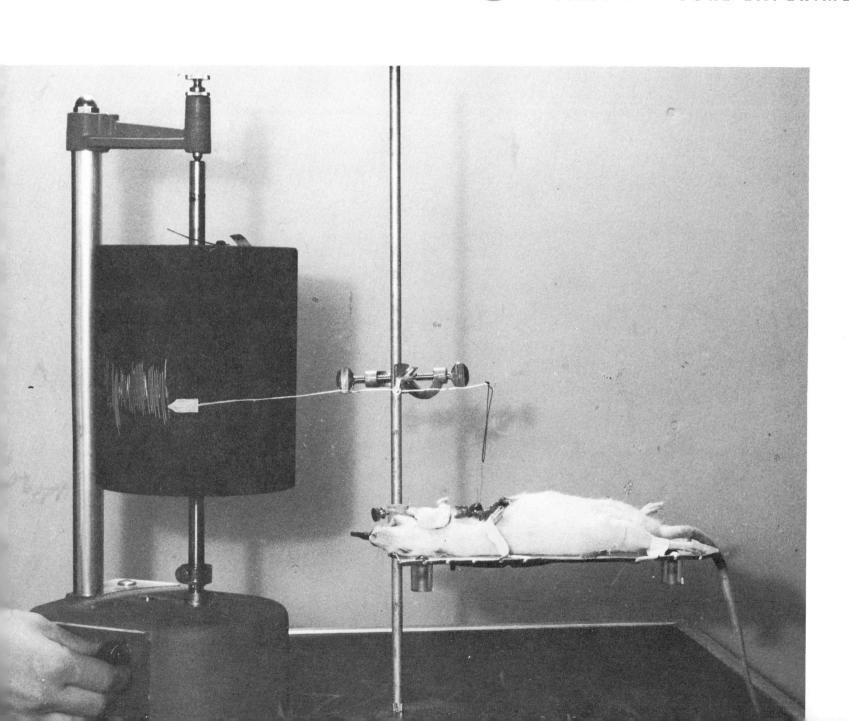

LIST OF EXPERIMENTS Part I

PRACTICE EXPERIMENTS
1. Handling, Weighing, Injecting, Anesthetizing, and Numbering Rats
2. Cannulation of Trachea, Carotid Artery, Jugular, and Femoral Vein

BLOOD
3. Determination of the Blood Volume
4. Red Count, White Count, and Differential Count
5. Hemoglobin Determinations
6. Coagulation Time, Density, Fragility, Sedimentation Rate, and Prothrombin Time

HEART AND CIRCULATION
7. Observations on the Heart in Situ; Effects of Heat, Cold, and Vagal Stimulation on the Heart Rate; Ventricular Fibrillation
8. General Directions for Blood Pressure Experiments; The Physiograph, Normal Blood Pressure; Effects of Adrenalin, Pitressin, Amyl Nitrite, and Chloroform on Blood Pressure
9. Effect of Nerve Stimulation on the Blood Pressure
10. Microscopic Observation of Circulatory Responses in Blood Vessels of the Mesentery
11. The Influence of the Carotid Sinus on the Blood Pressure
12. Effects of Changes in Blood Volume on the Blood Pressure

RESPIRATION
13. Respiratory Rate, Tidal Air, and Pulmonary Ventilation
14. Action of Carbon Dioxide on Respiration
15. The Hering-Breuer Reflex
16. Effect of Stimulating the Phrenic Nerve

DIGESTION AND METABOLISM
17. Effects of Vagal Stimulation on Gastric and Intestinal Motility
18. Effects of Vagal Stimulation and of Histamine on Gastric Secretion
19. The Secretion and Digestive Action of Saliva
20. The Secretion and Digestive Action of Pancreatic Juice
21. Effect of Decholin on Bile Flow

EXCRETION
22. Saline Diuresis
23. Reflex Contraction of the Bladder

ENDOCRINE SYSTEM
24. The Effects of Adrenalectomy and the Action of Cortin
25. The Effects of Thyroidectomy and of Thyroxin Administration on the Basal Metabolic Rate
26. The Effects of Ovariectomy and of Estrin Administration on the Estrus Cycle
27. The Effects of Castration and the Action of Testosterone
28. Mode of Action of Gonadotropic Hormones

NERVOUS SYSTEM
29. Muscle-Nerve Physiology
30. The Source of Energy for Muscular Contraction
31. The Bell-Magendie Law of Spinal Nerve Roots
32. Stimulation of the Motor Areas of the Cerebral Cortex
33. Effects of Autonomic Drugs on the Circulation
34. Effects of Autonomic Drugs on Gastric and Intestinal Motility

EXPERIMENT 1

HANDLING, WEIGHING, INJECTING, ANESTHETIZING, AND NUMBERING RATS

MATERIALS AND EQUIPMENT

Balance.

1-cc. hypodermic syringe and needles.

Warm, sterile, physiologic saline solution.

    Dissolve 0.9 gm. sodium chloride in 100 cc. distilled water, boil for several minutes.

Ear punch.

Scissors.

Special bent needle with ball tip for oral administration.

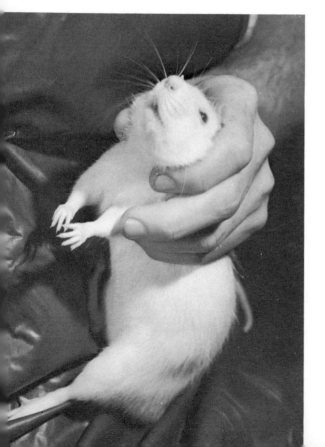

HANDLING

Rats not accustomed to handling may bite, but a rat cannot bite with his mouth shut. Therefore, hold him with the thumb and index finger around and below the lower jaw (not around the throat). Do not squeeze the body or throat; naturally, he will struggle to escape. It is easy to pick rats up by the tail, which, incidentally, does them no harm whatever, because they will pull away and can be picked up while straining. They cannot be held long suspended by their tails because they will climb up and bite. Gloves can be worn but are inconvenient and not necessary if proper technic is used.

Students unaccustomed to handling rats may well spend some time merely picking them up and transferring them from one box to another, until the "feel" of the rat and its reactions to manipulation have been thoroughly experienced and confidence is gained in the student's ability to control the animal.

continued

WEIGHING

A Hanson spring balance is satisfactory. Rats usually try to jump off the pan and escape; an easy method is to put the rat into a paper bag while weighing, later subtract the weight of bag.

INJECTING

The administration of substances in solution is frequently done in physiologic work. Three methods are here described: subcutaneous, intraperitoneal, and oral; intravenous and intra-arterial injections are described later. For parenteral or oral administration, it is not necessary to have the material at body temperature. It should, however, be warmed. Intravenous and intra-arterial injections must be made at body temperature (37°–38° C.).

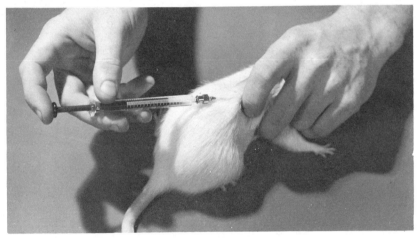

Subcutaneous injection

This work is done on a table. Grasp the rat firmly with the left hand, with one finger inclosing the lower jaw, not the throat. Pinch up the skin of the back between the thumb and the index finger. Holding the syringe in the right hand, pointing toward the head, insert the needle with a quick thrust and inject. Withdraw the needle and massage the site of injection to prevent escape of the fluid. If preferred, a towel can be placed over the rat to aid in holding it while the injection is made. Place the towel over the head and forefeet and hold the animal firmly.

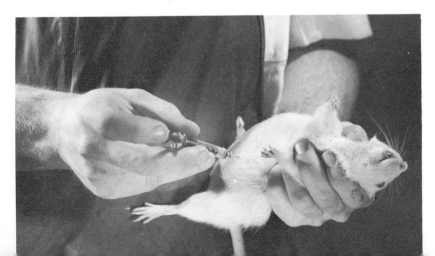

Intraperitoneal injection

Hold the rat in your left hand, with the thumb around the lower jaw (not the throat), and press the animal against the left side of your body. This holds the animal securely and tends to stretch the skin across the abdomen. Make the injection in the lower middle of the abdomen with a quick thrust. Use a short, ½-inch, 24–26-gauge needle.

Oral administration

Oral administration of liquid material is best made by using a large-gauge, bent syringe needle equipped with a ball tip. The rat is held in the left hand, with the sides of the head between the thumb and index finger, and the animal braced against the operator's left side. The syringe and needle having previously been filled with the desired amount of material, the ball tip is inserted into the animal's mouth and very gently pushed downward. It slides easily into the esophagus and is then pushed onward into the stomach. If any obstruction is felt, no force must be exerted, but another effort made to find the esophageal opening.

Practice injecting at first into lightly anesthetized rats until the technic is mastered, then repeat on conscious rats. Also practice controlling the volume injected, i.e., inject 0.25 cc., 0.50 cc., etc., from a full syringe. Sterile saline warmed to 30° or 35° C. is to be used in these practice injections.

ANESTHETIZING

As a basal anesthetic, Nembutal (sodium pentobarbital) is very satisfactory. The dose used is 50–60 mg. per kilogram for males, 40–50 for females, injected subcutaneously or intraperitoneally. Anesthesia is complete in 20 or 30 minutes and lasts about an hour if given subcutaneously; it sets in within 5 minutes if given intraperitoneally. If the operation is prolonged, a further dose of 20 mg/kg may then be given. If necessary, the Nembutal may be supplemented with small amounts of ether during the course of the operation.

For operations of short duration, ether alone may be used.

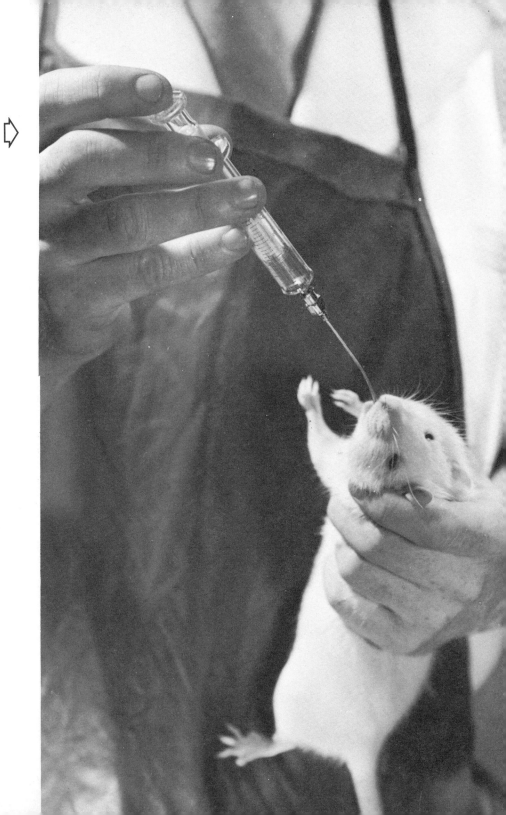

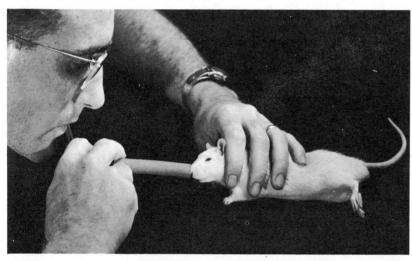

continued

A pad of cotton is sprinkled with ether and placed in a large jar; the cotton is covered with a folded paper towel, the rat is then placed in the jar, and the jar covered.

When the rat collapses, he is removed, and etherization is continued from a Gooch crucible containing a cotton pad moistened with ether. From time to time the ether is renewed by dipping the Gooch crucible into an evaporating dish containing ether.

Students should practice anesthetizing rats and observing the depth of anesthesia, as it progresses, by pinching the ears or tail. The respiratory movements are also a good index to the depth of anesthesia—when breathing is slow and shallow, too much ether is being given; when rapid and deep, too little.

RESUSCITATION

For the sake of practice, an overdose of ether should purposely be given and resuscitation attempted. (This is certain to happen occasionally during the course, and one should always be prepared for it.) Have at hand a soft rubber tube with bore large enough to fit over the rat's nose, with a glass tube attached. When respiration ceases, press the rubber tube against the rat's nose and gently blow and suck air from the lungs at a rate of about 30 per minute, pausing now and then to see whether the rat has resumed breathing. If too long a time has not elapsed, respiration can almost always be restored by this means.

NUMBERING

When the numbers are small, individual rats can be most easily marked for identification by rubbing methylene blue solu-

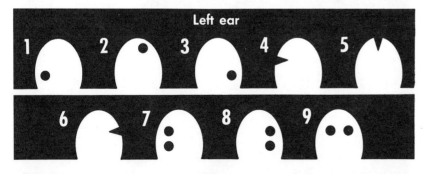

tion into the hair on the back of the head or various other parts of the body. For large numbers some type of numbering system must be employed. The system used in the accompanying chart is perhaps as simple as any. The holes are punched, and the nicks are made with scissors.

STUDENT'S REPORT

EXPERIMENT 2

CANNULATION OF TRACHEA, CAROTID ARTERY, JUGULAR AND FEMORAL VEIN

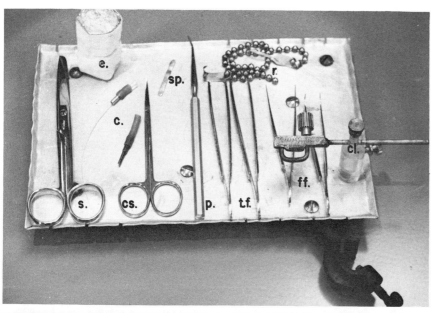

MATERIALS AND EQUIPMENT

Instruments

s.—general purpose surgical scissors, **cs.**—cannulating scissors (Clay Adams C1212), **p.**—blunt probe, **tf.**—tissue forceps (Clay Adams B-654/Cor SS, 4½ inches), **ff.**—eye forceps (Clay Adams C980/33 and C990/SS), **cl.**—carotid cannula clamp,* **r.**—retractors,* **sp.**—spatula,* **c.**—cannulae, and **e.**—etherizing, bulldog clamp (Phipps and Bird, 70-274), operating table.*

Cannulae.

Thread, fine and coarse.

Absorbent cotton.

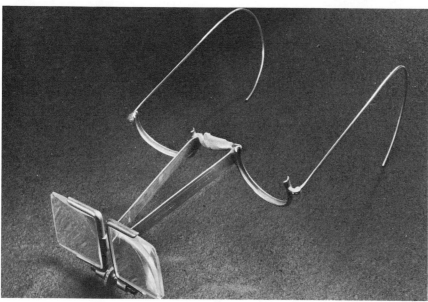

Jewelers' loupes, if desired.

Warm physiologic salt solution.

Bulldog clamp.

* All of these items can be made from common materials with minimum effort and only slight mechanical sophistication. (Continued on next page.)

THE OPERATING TABLE

Fasten a 9 × 6 sheet of aluminum ($\frac{1}{32}''$) with screws to a 4″ support ring with a standard support clamp so that the clamp will extend from what will be the front, right corner of the completed table. Steady the table by attaching Plexiglas feet ($\frac{5}{8}''$ rod, $\frac{3}{4}''$ long) to its undersurface at the other three corners with screws. Make a receptacle for receiving and holding the carotid clamp adapter and attach it to the front, right corner. It is made from 1″ long, $\frac{5}{8}''$ diameter Plexiglas rod by drilling a hole ($\frac{1}{4}''$ diam., $\frac{5}{8}''$ deep) into one end of the rod and then finished by drilling and tapping a second hole into the first near its top for a knurled, #8, set screw. This set screw and the one to be described on the carotid clamp adapter allow complete flexibility in positioning the clamp over the rat. Finish the table by turning up the edges of the aluminum sheet about $\frac{1}{4}''$ and by cutting notches along the edges to hold retractors.

RETRACTORS

Attach appropriately shaped, flattened and smoothed, stainless steel "hooks" to lengths of ball chain.

CAROTID CANNULA CLAMP AND ADAPTER

Drill and tap (#6) a hole in one end of a 1¼″ long, $\frac{3}{8}''$ diam. aluminum rod and screw a prethreaded, toilet valve, lift wire into it to make the side arm of the clamp. Drill two holes ($\frac{3}{16}''$) about ¾ inch apart through the rod. These will be used to hold and allow free passage of the lower jaw of the clamp. Make the lower jaw from a second lift wire by bending the wire twice into a U-shape. One limb of the U should be the threaded end of the lift wire. Cut off surplus wire; insert the lower jaw through the holes in the aluminum rod and attach a knurled nut (#6) to the threaded limb of the lower jaw.

An adapter is necessary for attaching the side arm of the clamp to the table and is made from 1¼″ long, $\frac{5}{8}''$ diameter Plexiglas rod. Turn one end of the rod on a lathe so that this end will slip to the bottom of receptacle on the operating table (slightly less than ¼″ diameter and $\frac{5}{8}''$ long). Drill a hole $\frac{5}{16}''$ deep in the opposite end of the rod and tap it for a set screw (#8, knurled head). Finish the adapter by drilling another hole ($\frac{7}{64}''$) through the rod intersecting the base of the previous hole at 90°. The side arm of the clamp will slip easily in this hole, but can be fixed firmly with the set screw.

SPATULA

Cut $\frac{3}{16}''$ wide strips from thin gauge stainless steel sheeting; cut it into 1½″ lengths; round both ends of each length, and polish all edges. Finish the spatula by bending a depression across it for cradling a vessel that is being cannulated. The notch should be about $\frac{3}{8}''$ from one end of the spatula.

In many of the experiments which follow, cannulation of blood vessels is necessary. This operation is somewhat difficult at first but becomes easy with practice. The carotid must be cannulated in blood pressure experiments and when large amounts of blood are to be obtained; the femoral or jugular, when giving intravenous injections.

CANNULAE

For the trachea a short (2 in.) piece of brass or glass tubing of the appropriate bore (depending on the size of the rat) may be used. Grooves are made along the brass tube, if glass is used the tip is flared, so that when the cannula is tied in place the tie will hold. The cannula is completed by attaching a short piece (1 in.) of rubber tubing to the end that will remain outside of the trachea.

In previous editions of this manual metal cannulae have been recommended for cannulating blood vessels, and even though these have been successfully used for years, cannulae prepared from polyethylene tubing have been found to be far more satisfactory.

Intramedic, PE 50, polyethylene tubing is the proper size tubing for cannulating larger vessels such as a carotid artery or a jugular vein in rats weighing between 120 and 400 grams. However, for cannulating smaller vessels, such as a ureter or a femoral vein, in the same sized animals, smaller PE 10 tubing is usually used.

The tubing can be connected to either a syringe or various pieces of apparatus with several types of adapters. Clay Adams' plastic tubing adapter (A-1025, size A) will accept both PE 50 and PE 10 tubing. Plans are included for a similar but much lighter and less awkward adapter in Figures 1 and 1a.

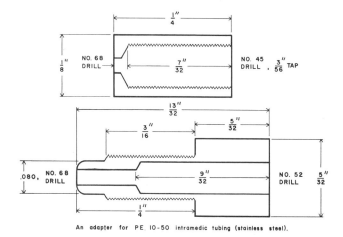

Fig. 1.—An adapter for PE. 10–50 intramedic tubing (stainless steel).

Fig. 1a.—Stainless steel adapter partially assembled.

A cannula can be assembled rapidly, using either of these adapters. A piece of the proper sized tubing is flared at one end by being held momentarily near the flame of a match. This end is tightly fixed by being placed in the forward end of the cannula adapter; the posterior end is then tightly screwed against it. The free end of the tubing is tapered sharply. The sharp point at the end of the tapered tip must be blunted. A short piece of ¼ inch rubber tubing is attached to the posterior end of the adapter and the cannula is completed by running a strengthening wire through the adapter and into the tubing. The strengthening wire must not protrude from the tip of the polyethylene tubing. The use of a strengthening wire of the proper diameter is essential (Nichrome wire: PE 50, 26 gauge; PE 10, 32 gauge) for it stiffens the tubing, insures a firm tie, and prevents crushing of the tubing.

PROCEDURES, CANNULATION OF THE TRACHEA, CAROTID ARTERY, AND FEMORAL VEIN

1 Anesthetize a rat with Nembutal. When anesthesia is complete, fasten the animal by each limb to either a table top or, preferably, to an operating table, with pieces of adhesive tape. Make an incision in the mid-line from slightly in back of the chin to the upper part of the thorax with a pair of general purpose scissors—never use fine tipped cannulating scissors for anything other than blood vessels or other fine structures. If the skin is elevated with a pair of tissue forceps prior to making the incision there is little danger of cutting into underlying structures. Next, expose the surgical field completely by grasping the skin on either side of the incision and pulling it laterally. The connective tissue sheath covering underlying muscles and, in the anterior region, the large paired submaxillary glands should now be apparent. Locate the submaxillary glands and divide and lay them to either side by holding the connective tissue sheath near the posterior free tip of the glands between two pairs of tissue forceps and then by pulling the forceps sideways and anteriorly. Be careful not to rip the large veins at the base of the submaxillary glands. A view similar to that in Figure 2 should be visible.

Cannulation of the Trachea

It is well to cannulate the trachea in any operation of long duration, as many respiratory difficulties are avoided by by-passing the nasal passages and the pharynx. Also, with a cannula in place, any fluid that accumulates in the trachea can be readily aspirated with a blunted hypodermic needle (17–19 gauge).

2 The first step in tracheal cannulation is to locate the sterno-

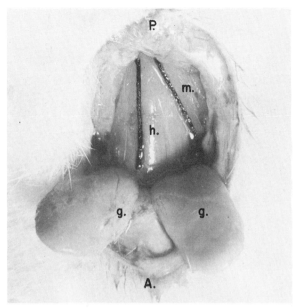

Fig. 2.—Tracheal Cannulation. **P.**—posterior, **A.**—anterior, **g.**—gland, **m.**—sternomastoideus, **h.**—sternohyoideus.

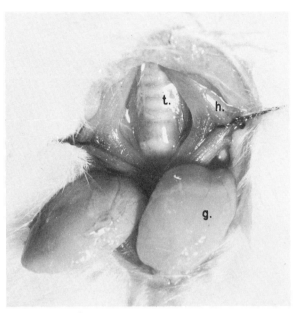

Fig. 3.—Tracheal Cannulation (cont.). **t.**—trachea, **g.**—gland, **h.**—sternohyoideus.

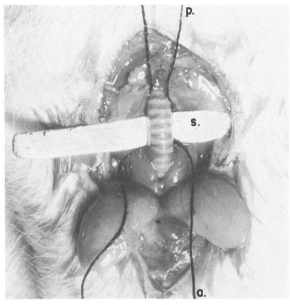

Fig. 4.—Tracheal Cannulation (cont.). **p.**—posterior, **a.**—anterior, **s.**—spatula.

hyoideus muscles (Fig. 2). The trachea lies under these and will become visible when they are separated along their central line with the closed tips of the curved forceps (Fig. 3). To isolate the trachea for cannulation, insert the closed tips of the curved forceps under it and separate the trachea from the underlying esophagus, by spreading the tips of the forceps so that the connective tissue holding these two structures together will be broken. Insert a spatula under the trachea and place two ligatures around the trachea as shown in Figure 4. Occasionally, bleeding may occur from blood vessels that pass along the lateral sides of the trachea. In order to stop or prevent this blood loss tie the anterior ligature

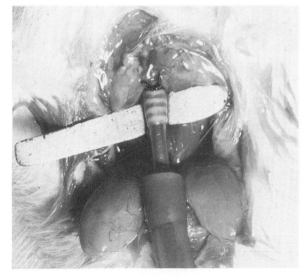

Fig. 5.—Tracheal Cannulation (cont.).

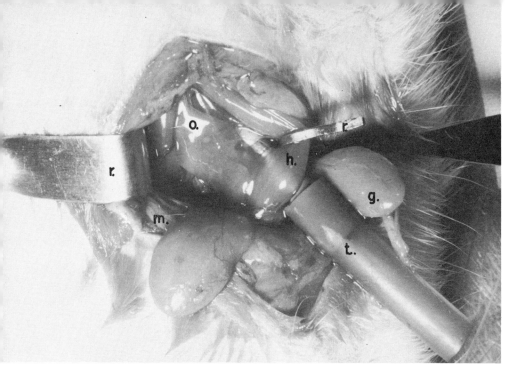

FIG. 6.—Carotid Cannulation. **r.**—retractor, **o.**—omohyoideus, **m.**—sternomastoideus, **h.**—sternohyoideus, **g.**—gland, **t.**—tracheal cannula.

FIG. 7.—Carotid Cannulation (cont.). **o.**—omohyoideus.

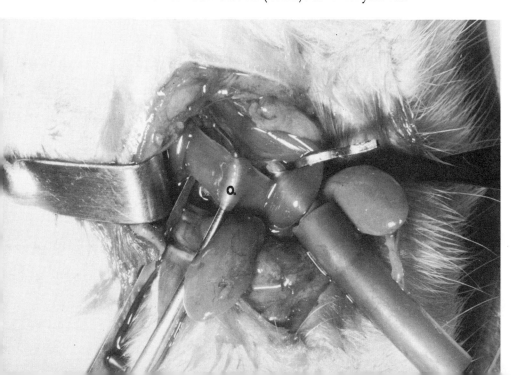

2 continued

firmly enough to occlude the vessels but not tightly enough to collapse the trachea.

3 Cut an opening in the trachea with the general purpose scissors and insert and tie the tracheal cannula in place with the posterior ligature. Test to see that the cannula is held firmly in the trachea by pulling on the cannula gently. If the cannula slips the ligature has not been tied over one of the notches on the cannula and the cannula should be removed and re-tied (Fig. 5). This and all ligatures should be tied firmly with a square knot and the loose ends from a tied ligature should be cut close to the knot and discarded immediately to prevent an unnecessary confusion of threads in the small surgical field.

Cannulation of the Carotid Artery

4 The carotid artery lies in a sheath beside the trachea between the sternohyoideus (h) and sternomastoideus (m) muscles and beneath the omohyoideus (o) muscle (Figs. 2, 6). Retract the sternohyoideus toward the midline and retract the sternomastoideus laterally to reveal the omohyoideus as in Figure 6. At this stage the white vagus nerve and the pulsating carotid artery are usually visible under the thin omohyoideus muscle. Cut this muscle after elevating it with the curved forceps (Fig. 7).

5 The carotid sheath, containing the carotid artery and the vagus and cervical sympathetic nerves, should now be obvious (Fig. 8). Separate and elevate the sheath and its contents from underlying tissue first by forcing the tips of the closed, curved forceps under it and then by letting the tips slowly separate.

6 Before the artery can be cannulated the connective tissue sheath surrounding it and the two nerves that lie beside it must be removed. To accomplish this slip the tips of a pair of straight, eye forceps under the carotid sheath and use the curved forceps to separate the nerves and the sheath from the blood vessel. If the nerves are to be used later they should be isolated with a loose tie at this time, otherwise drop them into the cavity. Check to see that the artery is completely free of superficial connective tissue; place a spatula under the artery. (Note: If surgery in other regions is to be done isolate the carotid with a loose ligature and cannulate it later; once a vessel is cannulated the animal must be heparinized.) Position anterior and posterior ligatures as shown in Figure 9. Tie the anterior ligature as far anteriorly as possible—this ligature will prevent a back flow of blood from the circle of Willis. Next, clamp the posterior end of the vessel with a bulldog clamp posteriorly to the ligature that will be used to tie the cannula in place. Place the clamp laterally, instead of vertically, and as far posteriorly as possible so that a maximum length of artery will be exposed for cannulation.

7 The artery is ready to be cannulated. Make a snip close to the headward tie with the fine pointed scissors and guide the cannula into the opening in the carotid with one hand while holding the spatula under the snipped portion of the artery with the other hand. Practice manipulating the vessel with the spatula so that the opening in the artery can be oriented for the easy insertion of the cannula. Tie the cannula firmly in place with the posterior ligature and make certain that the tie is made over a portion of the polyethylene tubing

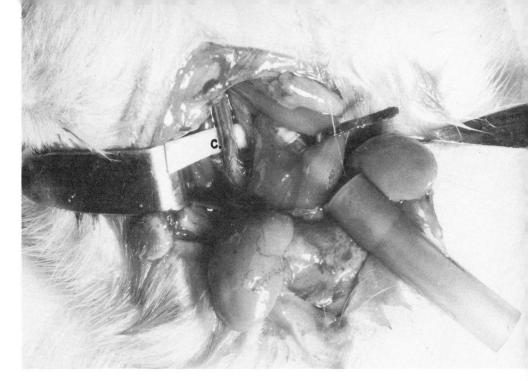

FIG. 8.—Carotid Cannulation (cont.). **c.**—carotid sheath.

FIG. 9.—Carotid Cannulation (cont.) **a.**—anterior, **p.**—posterior, **v.**—vagus nerve, **ca.**—carotid artery, **s.**—spatula.

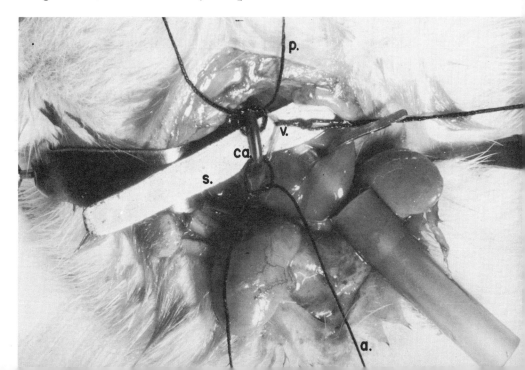

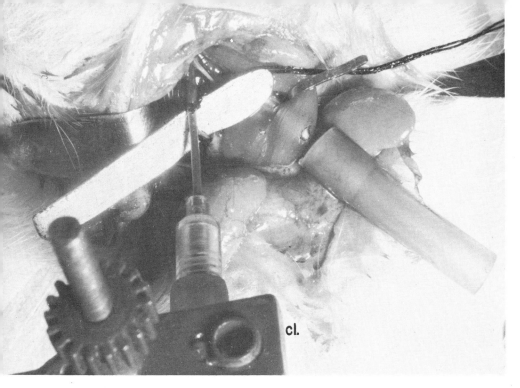

Fig. 10.—Carotid Cannulation (cont.). **cl.**—clamp.

Fig. 11.—Femoral Vein Cannulation.

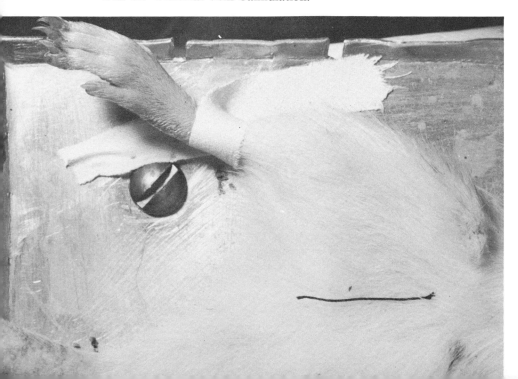

2 continued

containing the strengthening wire. If this is not done, the cannula might slip out of the vessel later or be crushed while the tie is being made.

8 Remove the strengthening wire; clamp the rubber tubing at the end of the cannula with the carotid cannula clamp, and remove the bulldog clamp (Fig. 10). (If a carotid cannula clamp is not available, occlude the rubber tubing with a heavy bulldog clamp or with a plastic-rod-plug.) A bulldog clamp should never be left on the carotid longer than necessary for the vessel in back of the clamp will deteriorate rapidly without a supply of blood.

9 If the animal is to be heparinized this should be done at this time. Fill a syringe with 0.2 ml. of heparin (1,000 units/ml) and attach the syringe to the free end of the rubber tubing on the cannula. Loosen the carotid cannula clamp until blood flows into and mixes with the heparin in the syringe. Slowly inject this mixture into the rat but take care not to inject any air into the vascular system. Withdraw and reintroduce blood several times to insure that all of the heparin has been washed out of the syringe.

Cannulation of the femoral vein

10 Expose the femoral area by first making a small incision in the skin along the inner surface of the thigh between the knee and the body (Fig. 11). Open the incision by inserting an index finger into it and rupture the connective tissue, which adheres the skin to the muscles of the leg, all the way to the

body wall. After retracting the body wall a view similar to Figure 13 should be seen.

11 Locate the femoral artery, vein, and nerves following the femur along the thigh. Isolate the vein by, first, removing the fascia that lies over it. Next roll the vein slightly to one side and find the circumflex iliac vessels that pass laterally into the thigh from the undersurface of the vein and artery. It is necessary to locate these vessels before proceeding since they vary in position. After locating them insert the tips of the forceps under the vein either bodyward or toward to the circumflex iliac vessels so that a maximum length of the vein can be cleared for cannulation in the next step. Rupture the fascia between the vein and the artery by bringing the tips of the forceps up between them and allowing the tips to slowly separate. Insert a spatula under the isolated vein and place a bodyward and a toeward ligature under the vein as pictured in Figure 12.

12 The vein is now ready for cannulation. The opening of the vein and the insertion of the cannula will be greatly facilitated if the toeward ligature is not tied prior to snipping the vein. Snip the vein and immediately tie the toeward ligature. Insert the tip of the cannula into the opening in the vein and run the tip down the vessel about an inch so that it will terminate in the larger iliac vein. Tie the bodyward ligature around the inserted cannula and insure that the cannula will not slip out in handling by also tying the loose ends of the toeward ligature firmly down on it. Remove the strengthening wire

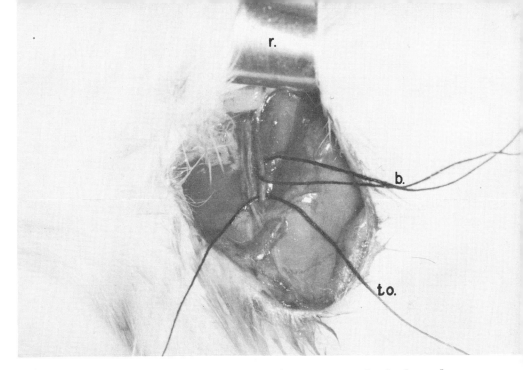

FIG. 12.—Femoral Vein Cannulation (cont.). **r.**—retractor, **b.**—bodyward tie, **to.**—toeward tie.

FIG. 13.—Femoral Vein Cannulation (cont.).

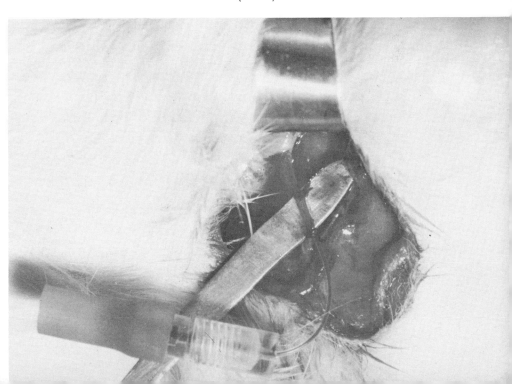

2 continued

and prevent bleeding from the free end of the cannula by plugging it with a short piece of plastic rod (Fig. 13).

13 The rat can be heparinized via the femoral cannula by injecting 0.2 ml. of heparin (1,000 units/ml) followed by 0.5 ml. of warm saline via the same syringe.

Cannulation of the Jugular Vein

Because of its close proximity to the heart, the jugular is not as desirous a site for making routine injections; but if a large volume of venous blood is wanted or if it is necessary to inject directly into the heart it should be used.

This is done in much the same manner as for the carotid. The jugular vein is located superficially, just under the skin, and more lateral to the trachea. Its walls are delicate, and it must be handled gently. It frequently collapses while being worked with, and, since it lies within a plexus of nerves, some reflex twitching of the animal occurs, despite the anesthetic.

14 Carefully dissect the vein free. For intravenous injection tie off the head end and clamp off the body end with a small bulldog. Make a preparatory loop, cut a small snip in the vein, and insert the jugular cannula, pointing toward the body. If the purpose is to obtain venous blood, these directions are reversed. Take care not to puncture the vein with the tip of the cannula.

15 Always clean out the cannulae thoroughly when they have been removed from the animal and be careful not to throw them out with the carcass. Wash instruments used and keep them in orderly array in a cabinet. Be especially careful with the fine-pointed forceps and scissors.

This experiment is essentially a practice exercise, and each student should have succeeded in performing three cannulations on the same animal (trachea, carotid, and a vein, preferably the femoral) within a reasonably short time and with the animal in good condition before proceeding with further experiments.

Note. A colored sound film showing the performance of all these cannulations is available and may be obtained by writing the Radio-Television Department of the University of Denver.

STUDENT'S REPORT

EXPERIMENT 3

DETERMINATION OF THE BLOOD VOLUME

MATERIALS AND EQUIPMENT

0.5 per cent solution of vital red dye.

Or use Evans Blue in a concentration of 0.3 per cent. The experiment given is with vital red.

Dissolve 100 mg. of the dye in 20 cc. of saline water.

Heparin.

There are a number of reliable commercial solutions on the market. (Heparin Lederle, 10 mg. per cubic centimeter, has been found very satisfactory.)

Instruments for carotid cannulation.
Colorimeter.
Centrifuge.

PREPARATION

Two rats are necessary for the determination. One is called the "control"; it merely supplies the plasma used in preparing the standard solution. The other is the experimental animal, which is to be injected with the dye solution. Its weight must be known to determine the amount of dye to be injected.

The quantity of dye to be injected is arrived at by a rough estimation of the animal's plasma volume. Repeated experiments have set the total blood volume of the rat at about 6 per cent of the body weight. If a rat weighs 300 gm., the blood volume will be about 18 cc. and the plasma volume about 9 cc. This figure will be used in making up the standard solution. For the dye injection, 1 cc. per kilogram is given, or 0.3 cc. for a 300-gm. rat.

PROCEDURE

1. Cannulate the carotid of the control.

2. Inject into the artery, by means of a 1-cc. syringe, directly through the cannula, an amount of saline equal to the volume of dye which will be injected into the experimental animal, plus 0.25 cc. of heparin. Mix the fluid remaining in the cannula with blood by repeatedly withdrawing about 0.5 cc. of blood into the syringe and then forcing it back.

3. Replace the 1-cc. syringe by a 10-cc. one and, after about 3 minutes, withdraw as much blood as possible, as rapidly as pos-

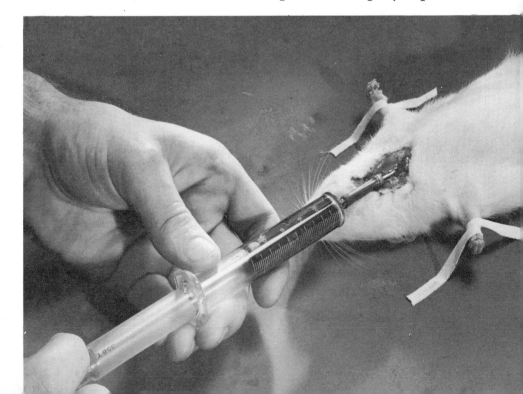

3 continued

sible—at least 5 cc.—and place directly in a centrifuge tube. Handle the blood carefully to avoid hemolysis.

4 Cannulate the carotid of the experimental animal in the same manner.

5 Inject 0.25 cc. of heparin plus the calculated amount of dye (1 cc. per kilogram) and mix as before. Note that the total amount of fluid injected into each animal is the same. After 3 minutes, withdraw as much blood as possible and place carefully in a centrifuge tube. Paste a label along the side of the centrifuge tube so that the level of cells and the level of the plasma may be marked after centrifuging. From this the relative volume of cells and plasma will be determined.

6 Centrifuge both samples for 30 minutes.

7 While the samples are in the centrifuge, make up the standard solution and prepare the colorimeter. The standard solution is made by adding the amount of dye injected to a volume of saline equal to the calculated plasma volume. That is, a 300-gm. rat would have an estimated plasma volume of 9 cc. and would have been injected with 0.3 cc. of the dye. The standard solution would then be 9 cc. of saline plus 0.3 of dye.

The colorimeter is prepared by adjusting each vernier to read zero when the cup is raised to rest against the bottom of the plunger. Both cups are then filled with the standard solution and set at 20 mm., and the light source is adjusted to give a uniform field. Remove the standard solution from the cups and save for future use. Wash and dry the cups.

8 Remove the samples from the centrifuge and mark the level of plasma and cells. Handle carefully to avoid mixing plasma and cells. The plasma of the control should be clear, that of the experimental animal should be stained pink.

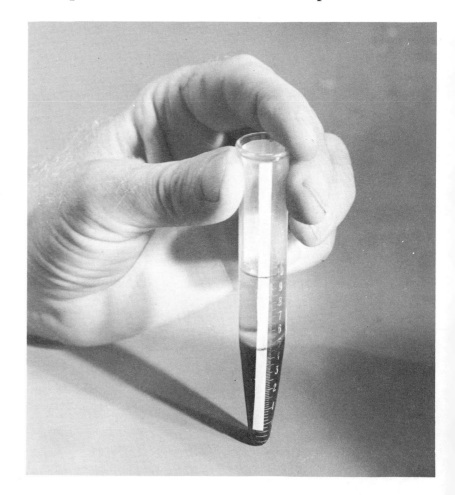

9 Take 2 cc. of plasma from the experimental sample and add to it 6 cc. of saline. From the control use 2 cc. of plasma plus 2 cc. of the standard solution plus 4 cc. of saline. The plasma from the control gives the standard solution the same degree of turbidity, when examined in the colorimeter, as that of the experimental solution. Syringes are the most convenient means of measuring.

Compare the two solutions in the colorimeter. Set the control solution at 20 and read the matching point of the other.

Now empty the centrifuge tubes and, by means of the levels you have marked, determine the proportions of blood cells and plasma. Measure from a syringe the amount of water required to fill the tube to the level of the red cells and from there to the top level of the plasma. Thus, if 2.5 cc. filled the tube to the first mark and another 2.5 cc. to the second mark, the total blood volume would be twice that of the plasma, or the hematocrit would be 50.

CALCULATIONS

Weight of rat: 380 gm.
Calculated plasma volume: $380 \times 0.03 = 11.4$ cc.
Volume of standard solution: 11.4 cc. saline + 0.38 cc. dye = 11.8 cc.
Colorimeter reading of experimental solution: 17.6; of standard: 20.
$\frac{11.8 \times 17.6}{20} = 10.4$ cc. of plasma.
Hematocrit: 50; $10.4 \times 2 = 20.8$ whole blood.
$\frac{20.8}{380} \times 100 = 5.4$ per cent.

Results obtained by students using this method usually run between 5.0 and 5.5 per cent. Published figures obtained by using various other methods range from 4.0 to 6.5 per cent.

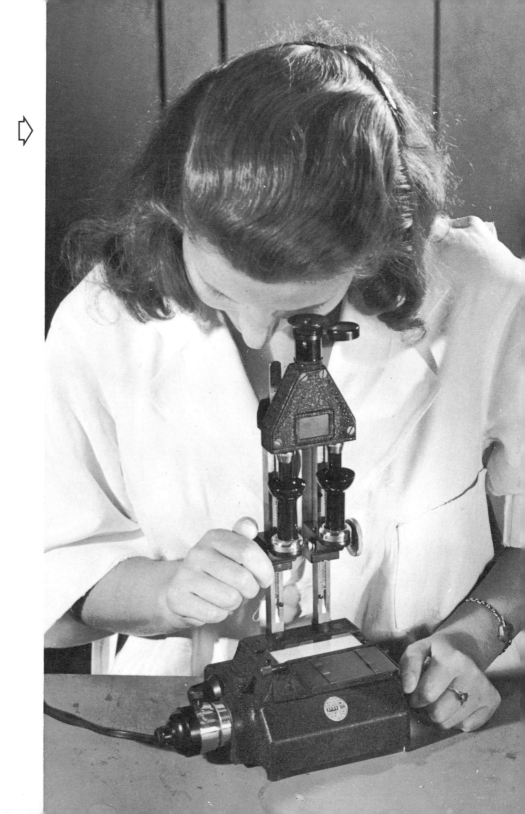

3 continued

STUDENT'S REPORT

EXPERIMENT 4

RED COUNT, WHITE COUNT, AND DIFFERENTIAL COUNT

MATERIALS AND EQUIPMENT

Hayem's red-cell diluting fluid.

    Dissolve 0.5 gm. mercuric chloride, 5.0 gm. sodium sulfate, and 1.0 gm. sodium chloride in 200 cc. distilled water. Filter before using.

White-cell diluting fluid.

    1 per cent acetic acid. Add 1 cc. glacial acetic acid to 100 cc. distilled water. Prepare fresh.

Wright's differential stain.

Spatula.

6 clean microscope slides.

Hemocytometer: counting chamber and blood pipettes.

Microscope.

Alcohol.

Scalpel or razor blade.

PROCEDURE

Obtaining blood sample

    Anesthetize the rat lightly with ether. Cleanse the tip of the tail with alcohol, permit alcohol to dry and snip off tip of tail

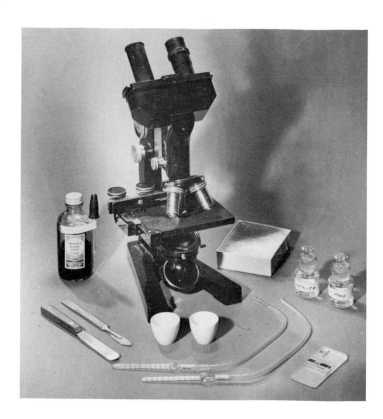
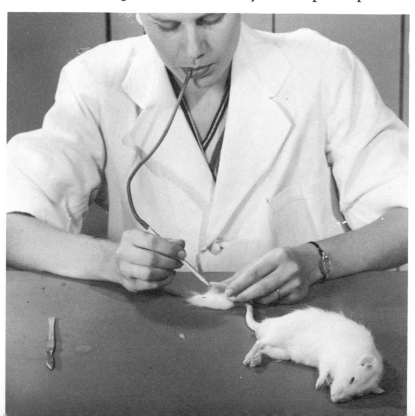

4 *continued*

cleanly with scalpel or razor blade. If the blood does not flow freely, do not squeeze the tail, as this will cause the tissue fluid to dilute the blood, but make another cut. Discard the first few drops of blood.

Blood for all three counts can be taken at once, therefore have the clean slides (see below for method of cleaning slides) and pipettes close by. After the samples have been taken, sear the rat's tail with a hot spatula to prevent further bleeding. After the cut has been made, make the smears for the differential count; as they dry, the other counts can be made.

DIFFERENTIAL COUNT

1 Wash several slides with soapy water. Rinse in alcohol, then ether, to remove all grease. Dry slides with soft towel which will leave no lint.

2 Place a small drop of blood about ¼ inch from the end of one of the clean slides. Do not allow tail to touch slide. Place another slide at a 45° angle just in front of the drop. Gently push the latter slide into the drop, allowing the blood to spread along the edge of the slide. Then pull the slide smoothly and not too rapidly to the end of the slide on which the smear is being made. Make several such smears and set aside to dry.

3 Cover the dry smear with Wright's stain, counting the number of drops. Allow to stand 1 minute; this fixes the smear to the slide. Now add an equal number of drops of distilled water to the stain. Mix. Let stand several minutes. Dry with filter paper.

4 Examine the smear under oil immersion and classify at least 100 cells. For identification of cells, refer to colored plates found in the textbook. Traverse the field thoroughly and cover a wide area. Calculate the percentage of each type of cell.

The usual range of percentages found in the rat is:

Lymphocytes	60–70 per cent
Neutrophils	30–35 per cent
Monocytes	0– 2 per cent
Eosinophils	0– 4 per cent
Basophils	0– 2 per cent

It will be noted that the proportion of lymphocytes to neutrophils is about the reverse of that found in the human.

RED COUNT

1 Hold the tail in one hand and with the other hold the pipette horizontally in the drop of blood. Do not permit the pipette to touch the tail. Draw the blood up to the 0.5 mark and wipe off any excess blood from the tip. Fill the pipette to the 101 mark with Hayem's solution, which should be kept at hand in

Preparation of blood smear for differential count.

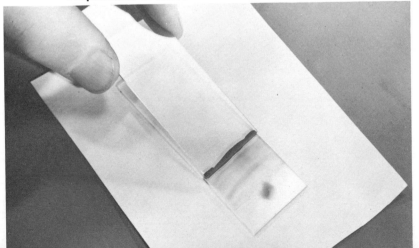

a convenient container. The measuring and diluting should be done with the greatest possible accuracy. Remove the rubber tube from the pipette, and, holding the thumb over one end and a finger over the other, mix contents very thoroughly. The count should be made as soon as possible after the dilution. If it is necessary for the diluted sample to stand, again mix before counting, as the cells have a tendency to settle.

2 Blow out the first few drops of diluted blood before filling the chamber. Be sure the chamber and cover slip are clean and dry. Touch a small drop to the edge of the cover slip which has been placed over the chamber, so that the fluid runs under the slip, filling the chamber. The Spencer Bright Line chamber is recommended.

3 Place the chamber under the microscope. It is best to locate the field under low power and then shift to high for the count. Count the cells in the four corner fields of 16 small squares each, including those touching the upper and left lines in each square. Calculate the average number of cells in each small square.

4 The dimensions of the chamber are the basis for calculation. Each of the smallest squares is $1/20 \times 1/20$ mm. The depth of the chamber is $1/10$ mm. If the pipette had been filled with blood to the 1 mark, the dilution would have been 1:100; however, it was filled to the 0.5 mark, and the dilution is therefore 1:200. The volume occupied by the fluid in a square is $1/20 \times 1/20 \times 1/10 = 1/4,000$ cu. mm. If the average number of cells

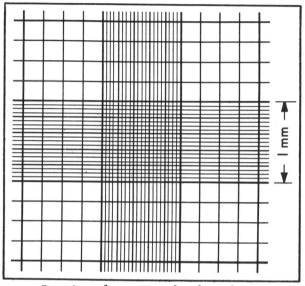

Drawing of counting chamber, showing spacing and dimension of squares.

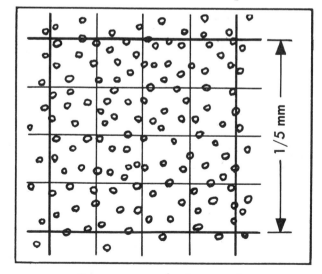

Appearance of microscopic field at magnification of 450 ×.

4 continued

were 10, there would therefore be 40,000 cells per cubic millimeter of diluted blood. Since the dilution was 1:200, the number of cells per cubic millimeter of whole blood would be $40,000 \times 200 = 8,000,000$.

Blood must not remain in the pipette very long, or the stem will become clogged. To clean, wash thoroughly with water, then dry with alcohol and ether. Wash the counting chamber with water and dry carefully with a soft towel or lens paper.

The red count in the rat is somewhat higher than in the human, running between 8,000,000 and 9,000,000 per cubic millimeter.

WHITE COUNT

1 The technic of taking the sample is the same as for the red count. The pipette, however, has a larger bore, and, when filled with blood to the 1 mark, the dilution is 1:10.

2 Draw the blood up to the 1 mark and dilute with 1 per cent acetic acid. Mix thoroughly and discard the first few drops.

3 Place the fluid in the counting chamber and count under low power. Count four of the largest squares and determine the average number of cells per square.

Since the area of the large squares is 1 sq. mm., the depth 1/10 mm., and the dilution 1:10, this average multiplied by 100 gives the number of white cells per cubic millimeter of blood.

The white count of the rat, as is true of the human, varies more than does the red count. The range is from 6,000 to 15,000, the average around 9,000 per cubic millimeter.

STUDENT'S REPORT

EXPERIMENT 5

HEMOGLOBIN DETERMINATIONS

The Hellige method of determining the hemoglobin content of blood is described here. The Hellige hemometer is used, in which the color of the acid hematin produced is compared with permanent standards. An alternative and more accurate method, the Van Slyke method, determines the hemoglobin on the basis of the volume of oxygen with which it combines. This method is described in Part II, Experiment 1.

HELLIGE METHOD

MATERIALS AND EQUIPMENT

Hellige hemometer.
  The new model (No. 305) is very satisfactory.
Scalpel or razor blade.
N/10 HCl.
  Dilute 1 cc. conc. HCl to 100 cc. with water.

PROCEDURE

1. Obtain a sample of tail blood in the manner previously described, drawing the blood into the pipette which is included with the hemometer to the point marked *1*. Wipe any excess blood from the tip of the pipette and dilute to the mark above the bulb with N/10 HCl. Note the time. Mix thoroughly and transfer the whole pipette content to the rectangular cell and place in the hemometer.

2. Five minutes after the mixture has been prepared, make a color comparison by revolving the disk and bringing one standard after another into view. When a good match is obtained, read the result in the rectangle on the front of the instrument.

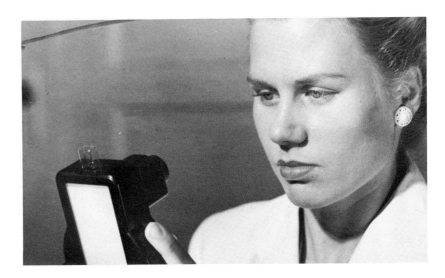

The result is given directly in grams of hemoglobin per 100 cc. of blood. Do not look at the colors more than a few seconds at a time and allow the eyes to rest frequently.

3. Clean the pipette in the same manner that blood-counting pipettes are cleaned and do so as soon as the test is completed. Also wash out the cell immediately. If this is not done, a film of acid hematin will form and cause high readings. If this film should be formed, it may be removed by soaking in a saturated sodium bicarbonate solution.

EXPERIMENT 6

COAGULATION TIME, DENSITY, FRAGILITY, SEDIMENTATION RATE, AND PROTHROMBIN TIME

The first two determinations can be made with tail blood, the others require a cannulation.

● COAGULATION TIME

MATERIALS AND EQUIPMENT

Capillary tube, 1-mm. bore and about 10 cm. long.
    The tube is scratched with a file at intervals of about 1 cm.
Stop watch.

PROCEDURE

1. Cut off the tip of the tail and discard the first few drops of blood. As a fresh drop appears, start the stop watch.

2. Hold the tube horizontally in the drop; capillary action will fill the tube with blood.

3. After the first minute, break off sections of the tube once every 15 seconds. At first the blood column breaks cleanly, but, when coagulation occurs, fine fibrin threads pull out from the broken ends. Normal coagulation time for rats is about 2.5 minutes.

● BLOOD DENSITY

MATERIALS AND EQUIPMENT

Benzene.

Chloroform.

Test tube.

Pipette.

Stirring rod.

PROCEDURE

1. Prepare a mixture of benzene and chloroform which will have a density of 1.06, which is near the density of blood. Place exactly 10 cc. of the mixture in a test tube or graduated cylinder.

2. Draw a small drop of tail blood into a fine pipette and dip the pipette into the center of the mixture. Gently force the drop out of the pipette and observe whether it rises or falls. If the former, add benzene, 0.1 cc. at a time; if the latter, add chloroform, keeping a careful record of the amounts added, until the point is reached at which the drop remains stationary. The density of the mixture is then the same as that of the blood.

CALCULATIONS

To prepare the benzene-chloroform mixture:
Let x represent the cubic centimeters of benzene; on a basis of 10 cc., 10 minus x equal the cubic centimeters of chloroform. Then:
$$0.89(x) + 1.5(10-x) = 1.06(10).$$

Therefore, 7.2 cc. of benzene plus 2.8 cc. of chloroform would give a solution whose density is 1.06. Prepare 20 cc. of such a mixture.

RESULTS OF TYPICAL EXPERIMENT

The droplet remained stationary after 0.2 cc. of benzene had been added to 10 cc. of the original mixture.

$$7.4 \text{ cc. benzene} \times 0.89 = 6.58 \text{ gm.}$$
$$2.8 \text{ cc. chloroform} \times 1.5 = 4.2 \text{ gm.}$$
$$10.2 \text{ cc. mixture} = 10.78 \text{ gm.}$$
$$1 \text{ cc.} = \frac{10.78}{10.2} = 1.057.$$

● FRAGILITY TEST

This test measures the resistance of red cells to hemolysis in hypotonic solutions.

MATERIALS AND EQUIPMENT

8 small test tubes.

PROCEDURE

1 Prepare sodium chloride solutions having concentrations of from 0.2 to 0.6 per cent, at 0.05 per cent intervals (0.2, 0.25, 0.3 per cent, etc.). Place about 2 cc. of each solution into small test tubes.

2 Cannulate the carotid and withdraw 5 cc. of blood. This will be ample for this and the following experiment.

3 Place 2 drops of blood in each tube and mix gently by inverting. Allow to stand at room temperature for 2 hours. The cells will have settled to the bottom, and hemolysis is indicated by the color of the fluid. If pinkish, with some cells at the bottom, hemolysis is partial; if red, with no cells, it is complete.

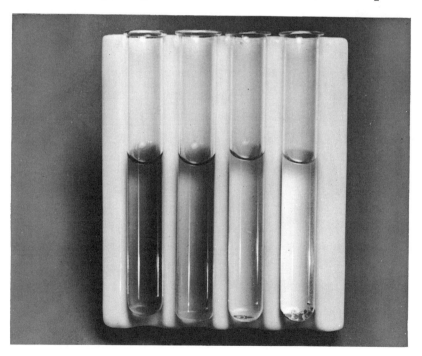

RESULTS OF TYPICAL EXPERIMENT

Complete hemolysis at 0.35 per cent and below.
Partial hemolysis at 0.4 and 0.45 per cent.
No hemolysis at 0.5 per cent and above.

6 continued

SEDIMENTATION RATE

MATERIALS AND EQUIPMENT

Blood sedimentation tube (Hellige, inside diam., 2.5 mm.; Cutler, inside diam., 5 mm.).

3 per cent sodium citrate solution.

PROCEDURE

1. Dilute the blood, 4 parts to 1 part of 3 per cent sodium citrate solution.

2. The blood is drawn into the sedimentation tube and fixed in the vertical position. A special rack is made for this purpose, but the tube can be pressed into a paraffin block and held vertically by a clamp. Read the millimeters of clear plasma formed at the top of the column every hour.

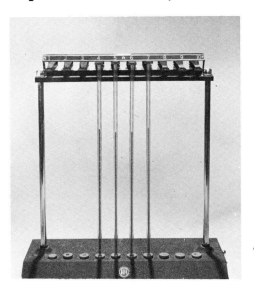

RESULTS OF TYPICAL EXPERIMENT

The rate will vary with the type of tube used, but in any case it is much slower in the rat than in the human. Using the Hellige tube, the average sedimentation rate in the rat is 0.5 mm. at the end of the first hour and 1 mm. in two hours.

PROTHROMBIN TIME

MATERIALS AND EQUIPMENT

1 per cent sodium oxalate in physiologic saline.

0.5 per cent calcium chloride solution.

4 small test tubes.

Centrifuge.

PROCEDURE

1. Cannulate the carotid but do not give heparin. Withdraw 4 cc. of blood into a syringe containing 0.5 cc. of 1 per cent sodium oxalate solution.

2. Centrifuge.

3. Place 5 drops of plasma in each of 4 small test tubes. To the first tube, add 2 drops of 0.5 per cent calcium chloride solution; to the second, 3 drops; to the third, 4 drops, and to the fourth, 5 drops. As the calcium chloride is added, mix by inverting and note the time. Observe for coagulation at intervals of 15 seconds until a tube can be inverted without disturbing the clot. This is called the "prothrombin time" and, in the rat, is between 3 and 3.5 minutes.

STUDENT'S REPORT

EXPERIMENT 7

OBSERVATIONS ON THE HEART IN SITU; EFFECTS OF HEAT, COLD, AND VAGAL STIMULATION ON THE HEART RATE; VENTRICULAR FIBRILLATION

MATERIALS AND EQUIPMENT

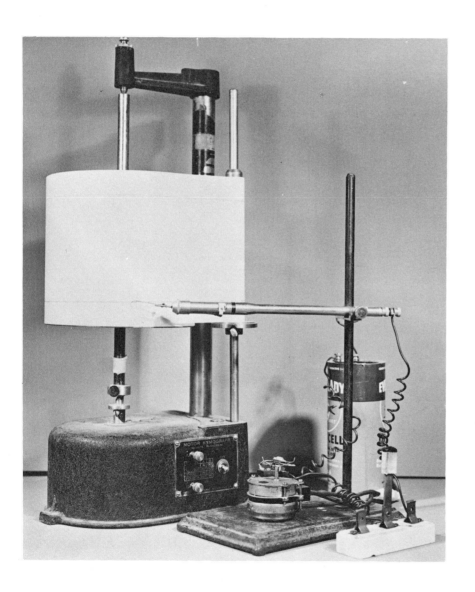

Timer and signal magnet.

The timer may be purchased, or an electric clock motor may be arranged to run a signal magnet at 1-second intervals. The model illustrated was made from a small 30-r.p.m. synchronous motor.

Heart lever.

The regular ink-writing yoke is used. One end of a thread is tied to the fulcrum, and a bent pin is attached to the other end, to be hooked into the heart. A small bulldog may be used as a counterweight.

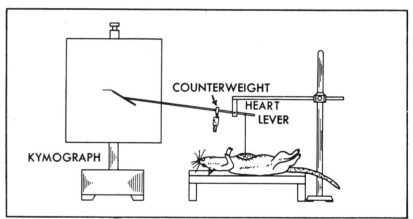

Attachment of heart to recording apparatus.

Kymograph.

The kymograph shown is the usual electric type, but modified for continuous recording by fitting an adjustable spool holder to the instrument rod. This type of kymograph is used throughout the course; where quantitative records are required, as in blood pressure experiments, ruled paper is necessary. Such paper may either be purchased or blank rolls may be ruled by setting up a battery of ink-writing pens and ruling the paper as it is unrolled from one shaft onto another, parallel, motor-driven shaft.

Artificial Respiration Apparatus

The above photograph illustrates a convenient and inexpensive type of artificial respiration apparatus. On it, a motor (20 r.p.m.) drives a bellows, as shown. (A cylinder and piston may be substituted for the bellows.) The outlet tube has a **T**-tube attached at the tracheal opening with a rubber tube and screw clamp attached. The volume of air going into the lungs may be controlled by adjusting the screw clamp. If the volume of the bellows is large compared to the lung volume, no valves are necessary.

Inductorium.

The inductorium consists of a primary coil and a secondary coil. When the circuit is made or broken by means of a switch or the interrupter, induced current flows in the secondary coil. The strength of the current flowing through the secondary coil is dependent upon the

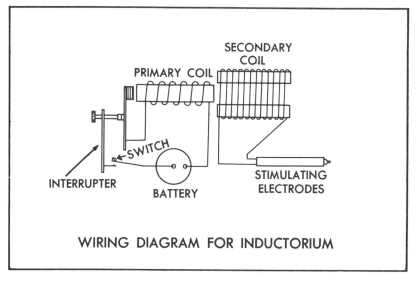

WIRING DIAGRAM FOR INDUCTORIUM

7 continued

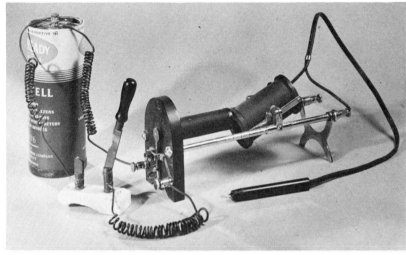

relative position of the two coils to one another. For a weak stimulus the secondary coil is placed so that only a short section is around the primary coil and for a very weak stimulus the secondary coil can be turned at an angle to the primary coil.

Kymograph.

Metal rod.

Hot and cold water.

Operating instruments.

PROCEDURE

1. Anesthetize the rat and cannulate the trachea preparatory to attaching it to the respirator. Free the right vagus nerve and pass a loop of thread around it so that it may easily be found later for stimulation.

2. Make a mid-line skin incision the length of the thorax and cross-incisions at either end, and retract the skin. Expose the

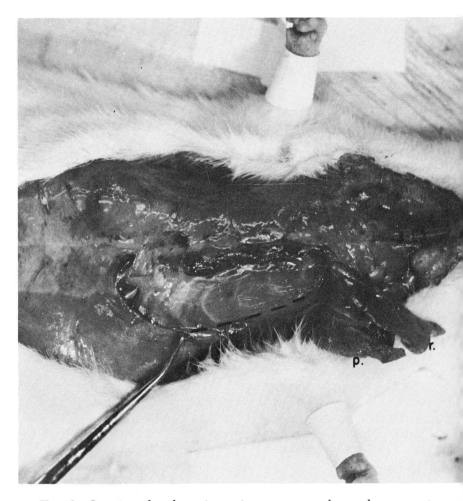

FIG. 1.—Opening the thoracic cavity. **p.**—pectoral muscles, **r.**—rectus abdominis, **t.**—tracheal cannula.

ribs by cutting the pectoralis muscles along the sternum, by cutting the rectus abdominis posteriorly, and then by displacing these muscles anteriorly as in Figure 1. Bleeding will occur but can be rapidly stopped by applying dry cotton swabs.

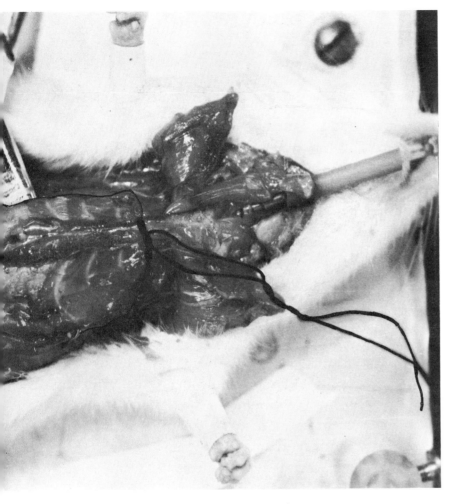

Fig. 2.—Opening the thoracic cavity (cont.).

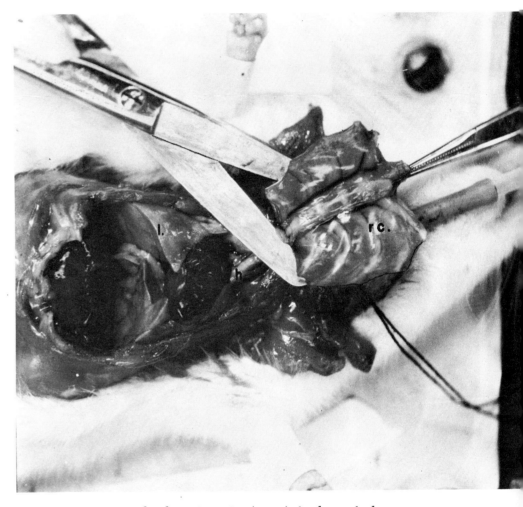

Fig. 3.—Opening the thoracic cavity (cont.). **h.**—heart, **l.**—lung, **rc.**—rib cage.

3 Cut the sixth through third ribs ½ inch lateral to the sternum. Put the animal on artificial respiration as soon as the first rib has been cut and adjust the respirator until the lungs fill normally. In cutting, be careful not to puncture the lungs with the scissors. (Note: incision line in Figure 1.)

4 Repeat the procedure on the other side. Now run a strong thread (#8) beneath the sternum at its upper end, between the second and third ribs. Tie tightly. The sternum may now be cut posterior to the tie and removed. The heart and lungs are now easily seen (Figures 2, 3).

7 continued

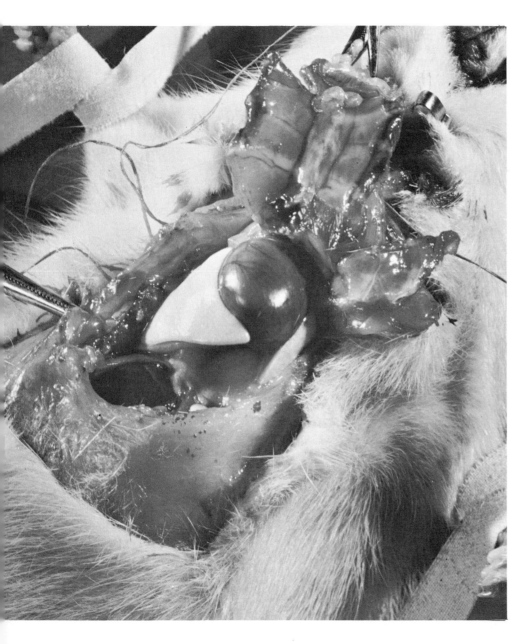

5 Notice the general appearance and action of the heart. Note the difference in color of the right and left atrium and the movements of systole and diastole, accompanied by a twisting motion. Observe and identify the venae cavae, the pulmonary artery, and the aorta.

6 Place the hook in the tip of the heart and arrange the apparatus so that the heart beat is recorded on the drum. By means of the timing record, the beats can be counted and the change of rate determined.

7 Having obtained a record of the normal beat, touch the upper part of the right auricle (region of the S.A. node) with the metal rod, which has been heated. Note the increase in rate. Let the beat return to normal and, having cooled the rod in cold water, repeat. The rate decreases.

8 Now return to the vagus. Tie the loop as far headward as possible and cut the nerve headward to the tie. Apply the electrodes from the inductorium to the peripheral end, making sure both electrodes are touching the nerve and using, to begin with, a strength of stimulus which can barely be felt on the tip of the tongue. Note slowing of the heart rate. Increase the strength of the stimulus sufficiently to stop the heart and continue to apply. Frequently, the heart will escape from the vagal inhibition and resume beating.

9 Increase etherization and then stop the respirator. Note that the blood in the left auricle becomes blue in color because of lack of oxygenation.

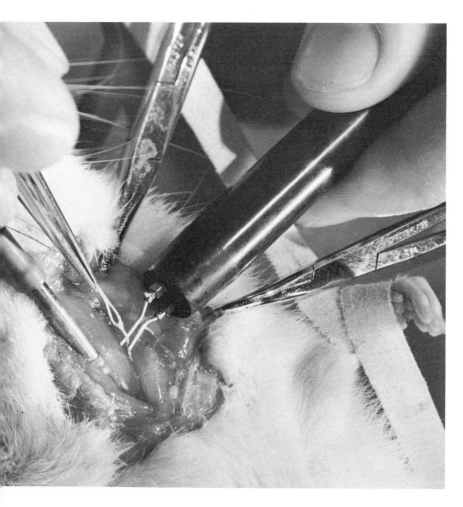

10 Resume artificial respiration. With the heart beating normally, apply the tips of the electrodes momentarily to the surface of the left ventricle, using a strong stimulus. The ventricle will go into fibrillation. The same effect will be produced by allowing the rat to breathe chloroform.

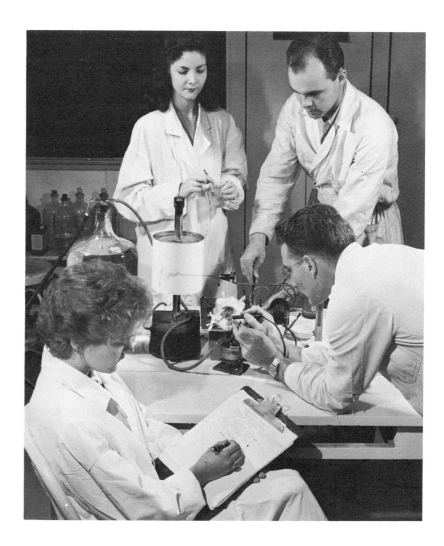

In this and all similar experiments, cotton pads soaked in warm saline solution should be freely used to prevent drying of the tissues, and loss of body temperature may be avoided by directing the lamp on the exposed regions.

7 continued

RESULTS—SAMPLE RECORDS

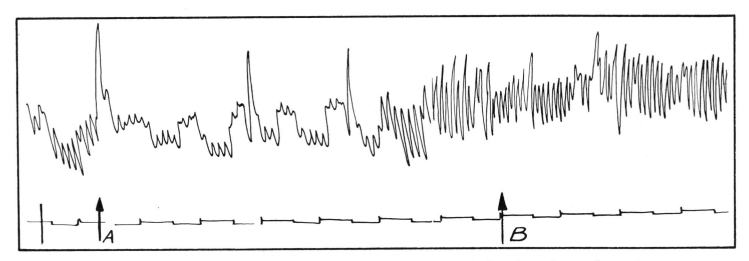

Effect of temperature on heart rate. At *A*, a cold rod was applied to the right auricle; at *B*, a heated rod. The recovery period following cold application has been cut out of the record.

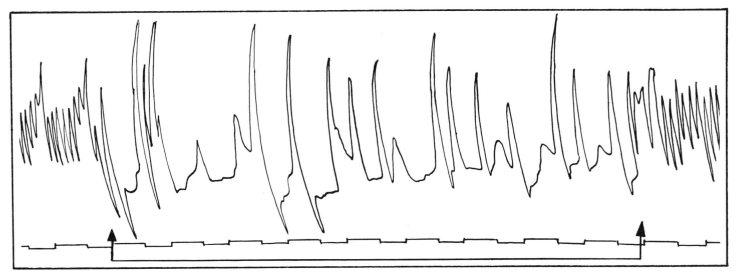

Vagal stimulation. Stimulus applied continuously between arrows. Break-through occurs periodically.

STUDENT'S REPORT

EXPERIMENT 8

GENERAL DIRECTIONS FOR BLOOD PRESSURE EXPERIMENTS; THE PHYSIOGRAPH; NORMAL BLOOD PRESSURE; EFFECTS OF ADRENALIN, PITRESSIN, AMYL NITRITE AND CHLOROFORM ON BLOOD PRESSURE

To obtain a recording of the blood pressure either of two pieces of apparatus may be employed: (1) a double-membrane tambour or (2) the physiograph. All records in this manual were obtained with the first apparatus; the physiograph is described below.

The double rubber membrane tambour is sensitive to the rapidly beating rat heart and is capable of producing pulse pressure tracings of between ½–1″ routinely. The tambour consists of a cylindrical chamber with a heavy rubber (fingertip from a surgical glove) membrane; above this, the upper opening is sealed by a thin (condom) rubber membrane. When the lower chamber is attached to the rat, variations in blood pressure are transmitted by the lower membrane to the upper membrane. Since the diameter of the upper membrane is small compared to the area of the lower, any movement of the latter will be greatly magnified.

The tambour should be adjusted as follows: Fill both chambers with saline; attach the tambour to a mercury manometer via the lateral limb on the T-tube connected to the lower chamber; clamp the other limb of the T-tube (this will be used later to attach the rat to the tambour). Gradually impose a pressure of 160 mmHg on the lower chamber with the syringe (glass) attached to the manometer while continually maintaining atmospheric pressure in the upper chamber by removing saline into the syringe (plastic) attached to this chamber. A pressure of 160 mmHg is sufficiently above average rat blood pressure to prevent blood from filling the lower chamber when it is opened to the rat's system. Atmospheric pressure is maintained in the upper chamber, since the upper membrane is most sensitive to changes in the lower chamber under these conditions.

Attach a rat to the tambour by connecting the carotid cannula to the lower limb of the T-tube and make certain that air is not introduced at this step. Remember that the presence of air anywhere in the recording system will diminish its sensitivity. Place the foot of the recording lever on the upper membrane; position the recording needle on the kymograph; cut off the manometer; and adjust pressure in the upper chamber and the position of the foot until an optimum curve is recorded upon the kymograph. The manometer can be opened to the system at any time and pressure changes noted on the drum. In fact, if the animal is in good condition and if an optimum pressure curve is not necessary, the manometer can be kept in the system continuously.

The double rubber membrane tambour and lever system can be purchased from Phipps and Bird Inc. or a chamber can be constructed from plastic.

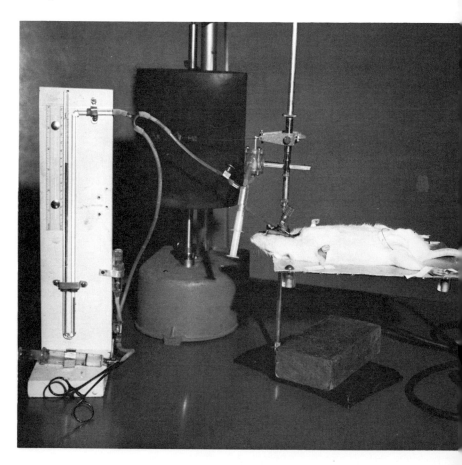

Arrangement of rat for blood pressure experiment

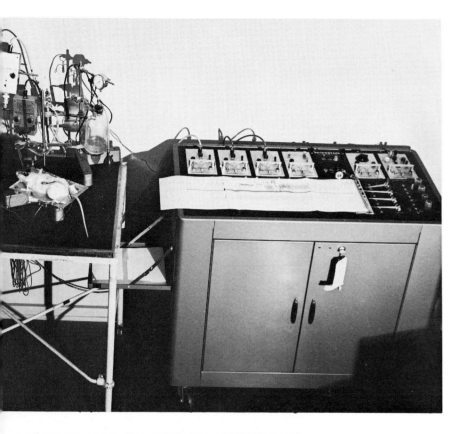

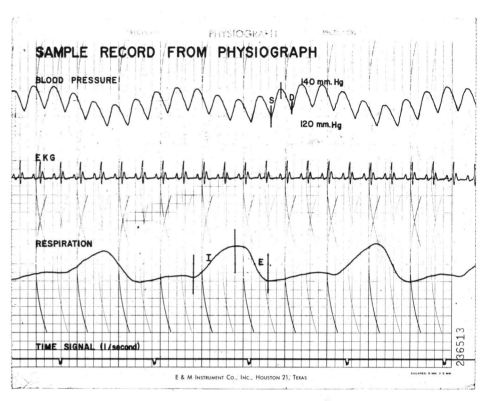

DESCRIPTION OF THE PHYSIOGRAPH

The physiograph is an electronic device for recording the measurable aspects of physiological activities such as blood pressure, the EKG, and respiration. It consists essentially of a pickup of some kind, usually a transducer, a conducting mechanism, and an amplifier that operates the recording pens. A three-channel physiograph is pictured and described. Each channel has the three elements mentioned above, so that three different physiologic events may be recorded simultaneously. In the case shown, the carotid is cannulated and hooked up to the transducer for blood pressure, electrodes, which pick up EKG signals and respiratory movements, are inserted into the skin. Each of these is recorded independently, as shown.

We have used the physiograph extensively at the University of Denver for several years and find it to have versatility and simplicity. It is described here with the thought that some schools will have or will purchase it and use it in this course. In any case, students should know something of modern methods of recording. Although all of the experiments listed in this Manual are described for using older methods the majority of these can be performed more accurately and more easily with an instrument like the physiograph. Also, since several variables can be recorded simultaneously a number of these experiments can be combined or elaborated upon to give the student deeper insight. The physiograph can be purchased from E & M Instrument Co., Inc., P.O. Box 14013, Houston, Texas 77021.

8 continued

DETERMINATION OF NORMAL BLOOD PRESSURE AND EFFECTS OF ADRENALIN, PITRESSIN, AMYL NITRITE, AND CHLOROFORM

MATERIALS AND EQUIPMENT

Blood pressure setup as described above (see also Part I, Experiment 33).

Adrenalin solution.
>Dilute ½ cc. 1:1,000 adrenalin to 100 cc. with saline, giving a solution containing 0.005 mg/cc.

Pitressin solution.
>Prepare a solution containing 0.1 pressor units per cubic centimeter. The commercial preparation (Parke-Davis) contains 20 pressor units per 1-cc. ampoule.

Amyl nitrite.
>"Pearls" of amyl nitrite (cotton-covered, glass ampoules, to be crushed between the fingers).

PROCEDURE

1 Prepare the animal and apparatus for a blood pressure tracing, cannulating trachea, femoral, and carotid.

2 Obtain a tracing of the normal blood pressure at different speeds of the drum. Note the individual pulse beats; by marking the drum with the timer, the heart rate can be determined. Note also (fast drum) that the rise in systole is steeper than the fall in diastole. The large waves seen in the tracing represent the alterations of blood pressure with respiration. Note that the rise occurs during expiration and the fall during inspiration and that the deeper the respiration, the greater the effect on the blood pressure.

3 Having obtained a normal tracing, inject, via the femoral, 0.005 mg. of adrenalin per kilogram of rat. (All intravenous injections should be made with the injected material heated to body temperature.) Note the sharp rise in pressure and its brief duration. The injection may be repeated with slightly smaller or larger doses.

4 After the effect of adrenalin has worn off, inject 0.1 pressor units of Pitressin per kilogram of rat. Note the slower rise and the much more prolonged action of Pitressin.

5 After the blood pressure has again returned to normal, crush a pearl of amyl nitrite between the fingers, drop into a small crucible or beaker, and hold the tracheal cannula over the vessel. An immediate sharp fall in pressure results. Remove the crucible at once; prompt recovery follows.

6 Finally, soak a bit of cotton with chloroform, place in a crucible, and let the animal breathe the vapor. A sharp fall in pressure results. The rat may be killed with chloroform, continuing the record until death.

RESULTS—SAMPLE RECORDS

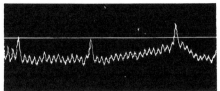 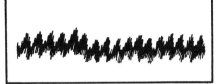  

Comparison of normal blood pressure records, using mercury manometer (*left*), with those obtained with double-membrane tambour. Note that with the mercury manometer the respiration waves, being slower, are visible, but the pulse waves can hardly be seen. Records taken with the tambour are made at different drum speeds.

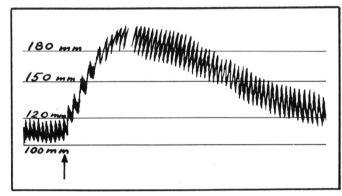 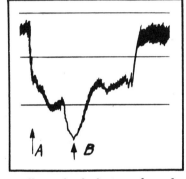 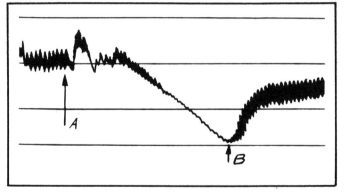

An injection of 0.005 mg/kg of adrenalin was given at point marked by arrow.

Effect of inhalation of amyl nitrite (*A*), and recovery following removal (*B*).

Effect of inhalation of chloroform (*A*), and recovery following removal (*B*).

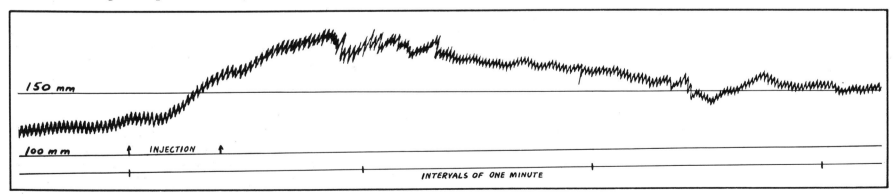

Effect of injection of 0.1 pressor units per kilogram of Pitressin.

8 continued

STUDENT'S REPORT

EXPERIMENT 9

EFFECT OF NERVE STIMULATION ON THE BLOOD PRESSURE

Four types of nerves are involved in blood pressure control—depressor and pressor nerves, which are afferents, and vasodilator and vasoconstrictor nerves, which are efferents. In this experiment the effect of stimulating a depressor nerve is shown, using the central vagus; of a pressor nerve, using the central femoral; of a vasodilator nerve, using the peripheral vagus; and of a vasoconstrictor nerve, using the peripheral splanchnic.

MATERIALS AND EQUIPMENT

Blood pressure setup as described in Part I, Experiment 8.

Inductorium.

Ringer-Locke solution.

> Dissolve 8.0 gm. sodium chloride, 0.2 gm. potassium chloride, 0.2 gm. calcium chloride, and 0.1 gm. sodium bicarbonate in 1 liter of distilled water.

PROCEDURE

1 Cannulate the left carotid artery, being careful not to injure the vagus while doing so. Gently free a portion of the right vagus and pass a loop of thread around it so that it can readily be found later. When working with nerves it is better to use warm Ringer-Locke solution to keep the tissue moist, rather than saline.

2 Make an incision over the groin and expose the femoral nerve, which will be found lying along the femoral artery. Place a loop around it for easy identification and close the incision with a hemostat.

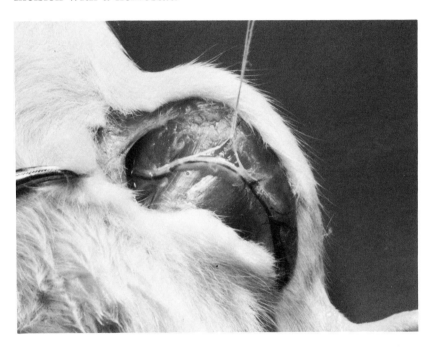

3 Connect the carotid cannula to the tambour. Take a normal blood pressure record. Return to the vagus and place two ties around it, close together, in its mid-portion. With the drum running, cut between the ties. Using the tie to hold up the end of the nerve, stimulate the central end with shocks of moderate intensity. In the rat, depressor fibers predominate, and the effect, with stimulation of moderate intensity, is a fall in blood

9 continued

pressure. Stimulate the peripheral end with low-intensity shock; there will result a pronounced fall in blood pressure.

4 The femoral nerve contains pain fibers, which are pressor in effect. Tie the loop around the femoral and cut the nerve peripheral to the tie. Holding the nerve by the tie, stimulate with current of moderate intensity; a rise in blood pressure results. The same effect may be demonstrated by lightening the anesthesia and pinching the ears. If this is planned, the dose of Nembutal given should be reduced to 40 mg/kg, or a different rat used which has been anesthetized with ether only.

5 The cervical sympathetic is a vasoconstrictor nerve but supplies relatively few vessels. It may be found instructive to free and stimulate its peripheral (headward) end. Usually no perceptible rise occurs, which is in contrast to the effect which will be obtained with the splanchnic.

For this part of the experiment another rat is to be used.

6 The nerve to be stimulated is the greater splanchnic, on the right side; it is better to locate the nerve before making the cannulation in order to avoid trouble with the carotid while moving the rat around. With the animal lying on its left side, make a lengthwise skin incision about 1 cm. to the right of the dorsal mid-line and extending from the lower ribs about halfway to the hip. Make transverse incisions at either end and reflect the skin.

7 The splanchnic lies deep in the angle formed by the pillar of the diaphragm and the ventral face of the psoas. To reach this region it is necessary to cut through the intervening sheets of thin muscle. These are the latissimus dorsi, sloping headward and ventrally, and the external oblique, sloping tailward and ventrally. Their dorsal attachment, the lumbodorsal aponeurosis, should be cut and the muscles reflected ventrally. A small muscle in the anterior region of the wound will now be seen, attached to the last rib, its fibers sloping dorsally. Detach this muscle from its attachment to the rib. The abdominal aponeurosis of the abdominal muscles are attached to the transverse processes of the lumbar vertebrae. These are now freed from their attachments.

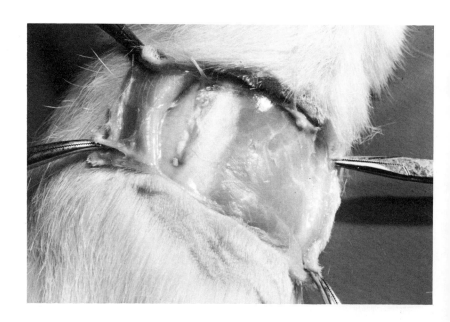

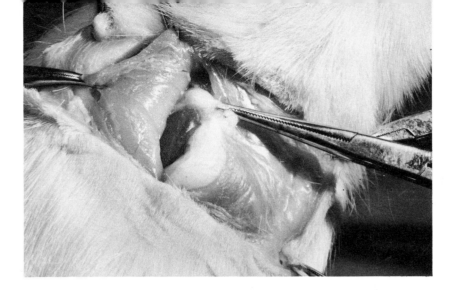

8 The ventral face of the vertebral column is covered with a thick, lengthwise-running muscle, the psoas. Gently moving the liver aside with the handle of a scalpel, explore its face ventrally and headward; an angle found between it and the right diaphragmatic pillar is exposed and, deep in it, emerging from the psoas and sloping toward the upper pole of the kidney, lies the splanchnic nerve. It is a small nerve, and only a short section of it can be seen; it disappears distally into the fat surrounding the kidney. Gently free as much of it as possible, using pledgets moistened with warm Ringer-Locke to remove blood, and pass a loop of thread around it. Close the incision with wound clips.

9 Now cannulate the right carotid and attach to the tambour in such a way that the animal can again be laid on its left side.

10 Returning to the splanchnic, tie the loop as far proximally (toward the vertebral column) as possible, leaving room to cut proximal to the tie. Using the tie to lift the nerve, stimulate with weak shocks from the inductorium, being sure that both electrodes are in contact with the nerve but not touching other tissue. A prompt rise in blood pressure results from the direct constrictor effect upon the arterioles, followed quickly by a second rise, accompanied by an increase in the rate and strength of the heart beat. These latter phenomena are due to the liberation of adrenalin, since the splanchnic nerve includes fibers running to the adrenal medulla.

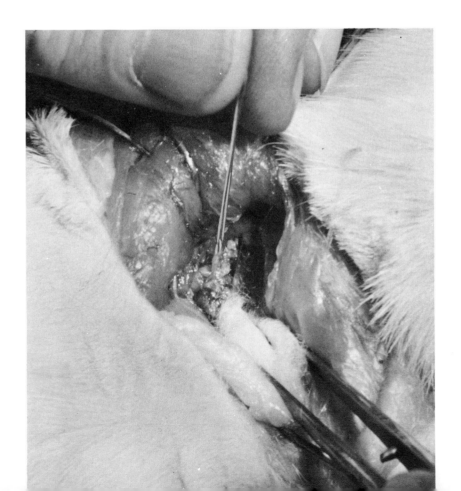

9 continued

RESULTS—SAMPLE RECORDS

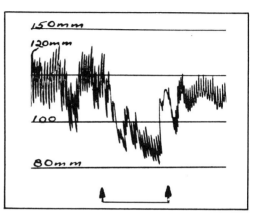

Effect of stimulation of central end of cut vagus. Stimulus applied between points marked by arrows.

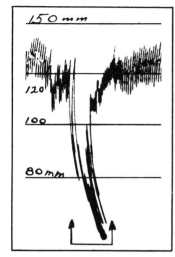

Stimulation of peripheral end of cut vagus.

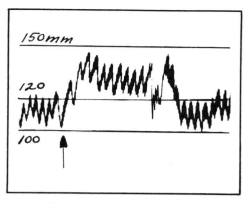

Effect of stimulation of central end of cut femoral nerve.

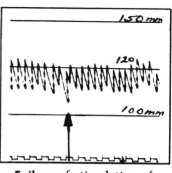

Failure of stimulation of central end of cut cervical sympathetic nerve to affect blood pressure.

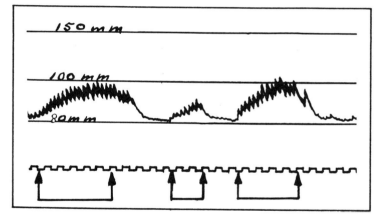

Effect on the blood pressure of stimulating the peripheral end of the splanchnic nerve.

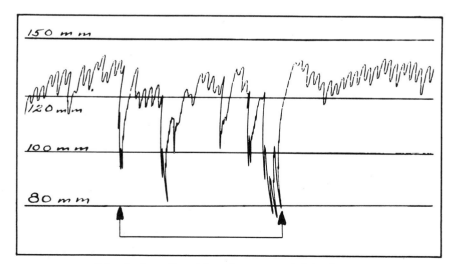

Stimulation of peripheral vagus with fast drum.

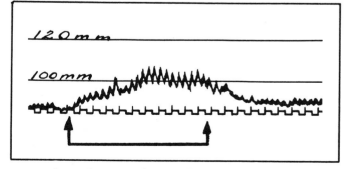

Secondary rise due to adrenalin liberation.

STUDENT'S REPORT

EXPERIMENT 10

MICROSCOPIC OBSERVATION OF CIRCULATORY RESPONSES IN BLOOD VESSELS OF THE MESENTERY

MATERIALS AND EQUIPMENT

Microscope.
Inductorium.
Adrenalin solution; 0.005 mg/cc.

PROCEDURE

1 Expose the right vagus nerve and place a loop around it but do not tie.

2 Cannulate the right femoral vein for adrenalin injection.

3 Make a short mid-line incision in the abdomen. On a platform built alongside the microscope and having the same height as the stage, lay the rat on its left side. Very gently pull a loop of the intestine across the stage so that the mesentery will lie across the light opening. Under low power examine the field and find a place from which a good view of the circulation can be obtained. Note the rapid blood flow through the smaller arteries; the slow, single-file progress of red cells through the capillaries; and the sluggish flow through the veins. Do not leave the viscera exposed for long periods; moisten with warm saline and return to the body cavity frequently.

4 While a good field is under observation, have the adrenalin injected via the femoral, giving a dose of 0.005 mg/kg. Note the almost immediate response of the arterioles—they will constrict, and the blood flow through the capillaries will stop. As the adrenalin effect wears off, the flow is resumed. This may be repeated several times so that all members of the group may observe the effect, but the intestine should be returned to the body cavity for a few minutes after each test.

5 Now tie off and cut the vagus. Having a favorable view, have the peripheral end stimulated. The effect is not always uniform, but usually at the instant of application there is a speeding-up of blood flow because of vasodilatation. As the stimulus is continued, however, the flow stops because of the severe fall in the general blood pressure.

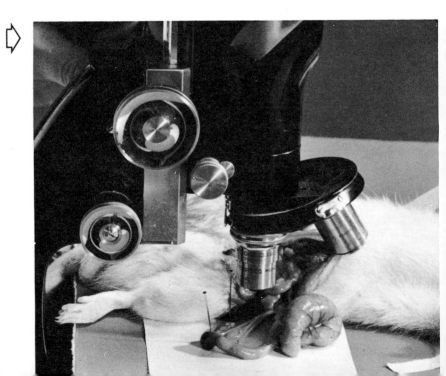

STUDENT'S REPORT

EXPERIMENT 11 — THE INFLUENCE OF THE CAROTID SINUS ON THE BLOOD PRESSURE

Sensory nerve endings lying within the wall of the carotid artery at its bifurcation (the carotid sinus) react to changes in pressure, an increase of pressure within the sinus producing a fall, and a decrease a rise, in the general blood pressure. By recording the blood pressure from one carotid and altering the pressure within the other, the operation of this mechanism may be shown.

MATERIALS AND EQUIPMENT
Blood pressure setup.

PROCEDURE

1 Cannulate the right carotid artery.

2 Pass a loop of thread under the left carotid below the bifurcation and another loop above the bifurcation, the latter inclosing both the internal and the external branches. Care must be taken in doing this not to disturb any of the nerve structures which are found in this region. Dissection should be kept to a minimum and the loop passed deeply.

3 The carotid cannula is now attached to the tambour, and a normal record is taken. Now exert traction on the lower loop; the pressure within the carotid is now reduced, as shown in a later experiment. The blood pressure as registered from the other carotid rises. The traction exerted to close the vessel should not be great enough to stretch the bifurcation; it is perhaps preferable to lift the artery by means of the loop and attach a bulldog.

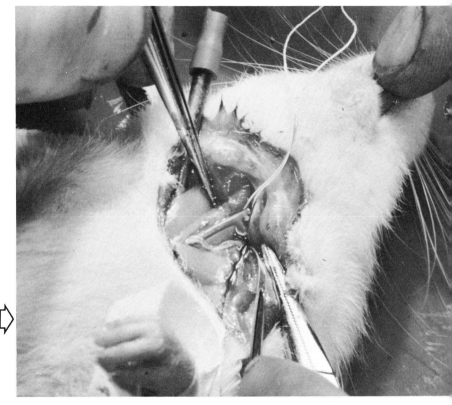

Dissection to show bifurcation of carotid. In performing this experiment, dissection should not be carried out as rapidly as shown.

4 Now do the same with the upper loop. This restricts the flow beyond the bifurcation, and a greater pressure is exerted within its walls. The blood pressure will be observed to fall. Repeat both parts of the experiment to be sure of the results. If a fast drum and timer are used, it is also possible to observe the effects of the sinus mechanism on the heart rate.

RESULTS—SAMPLE RECORDS

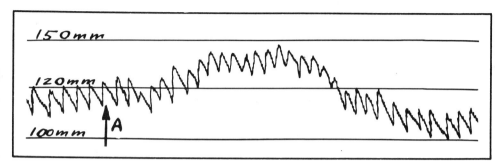

The body end of the carotid was occluded; the pressure in the sinus is thereby reduced, and the general blood pressure rises.

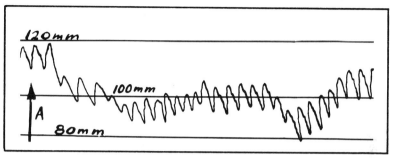

The head end of the carotid was occluded; the pressure in the sinus is thereby raised, and the general blood pressure falls.

STUDENT'S REPORT

EXPERIMENT 12

EFFECTS OF CHANGES IN BLOOD VOLUME ON THE BLOOD PRESSURE

This experiment may be modified in different ways; as here described, one rat is used to demonstrate the effects of reduction in blood volume, the other to show effects of increase in blood volume, on the blood pressure.

MATERIALS AND EQUIPMENT

Blood pressure setup.

PROCEDURE

1. Cannulate both carotids of the first rat. Heparinize.

2. Attach one cannula to the blood pressure apparatus.

3. By means of a syringe, slowly withdraw measured amounts of blood. Note that the rate of withdrawal affects the blood pressure drop and that recovery is prompt if only small amounts are withdrawn. Determine the volume for which the fall is relatively permanent and note that reinjection restores the blood pressure promptly.

4. Repeat the withdrawal of blood and replace with an equal amount of saline warmed to body temperature. Its immediate effect is the same as with blood, indicating that the fall was due to a reduction in the volume of circulating fluid. Kill the rat by withdrawing larger and larger amounts of blood, waiting after each withdrawal to see if recovery occurs. Save this blood for the second rat.

5. Cannulate left carotid artery and the right femoral vein of another rat. Heparinize. While taking a normal record, inject 1 cc. of saline warmed to body temperature via the femoral. An increase in blood pressure occurs, which is followed by a slow return to normal.

6. Now inject 1 cc. of the blood obtained from the first rat, which has been kept warm. A rise, this time usually of longer duration, follows.

RESULTS—SAMPLE RECORDS

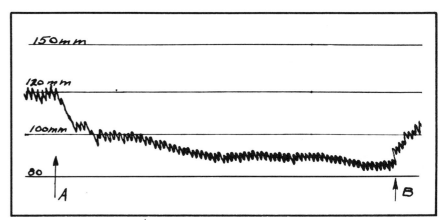

Effect of withdrawal of 1 cc. of blood at A and introduction of 1 cc. of saline at B.

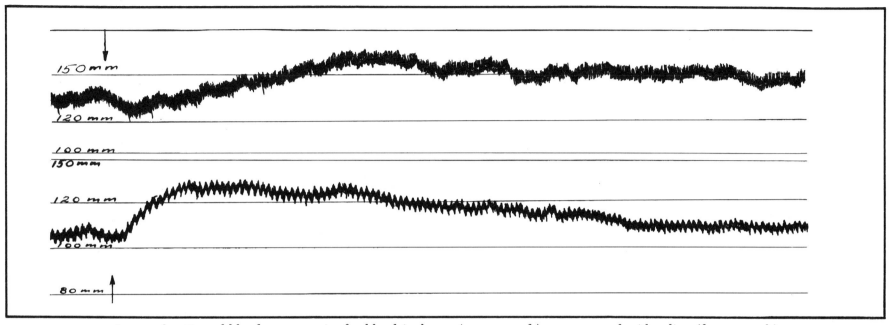

Longer duration of blood pressure rise for blood is shown (*upper graph*), as compared with saline (*lower graph*).

 continued

STUDENT'S REPORT

EXPERIMENT 13

RESPIRATORY RATE, TIDAL AIR, AND PULMONARY VENTILATION

MATERIALS AND EQUIPMENT

Pair of one-way valves.
Collecting bulbs.
    Leveling bulb is mounted on rack-and-pinion support.
Stop watch.

PROCEDURE

1. Anesthetize a rat and cannulate the trachea.

2. With the valves arranged as illustrated, connect the outlet valve to one of the collecting bulbs, maintaining a slight positive pressure (about 1 cm. of water) against the valve to prevent leakage. As air collects, lower the other bulb, maintaining the positive pressure. When the water in the collecting bulb reaches the upper calibration, start the stop watch. Count the respiration rate, which is easiest done by watching the inflow of air through the trap, and collect 100 cc. of air.

CALCULATIONS

Note the time required and from it calculate the pulmonary ventilation, i.e., the number of cubic centimeters of air breathed per minute. Divide this figure by the number of respirations per minute; this gives the tidal air, i.e., the number of cubic centimeters per breath.

Both the respiratory rate and the pulmonary ventilation vary greatly, depending upon the size of the rat and the depth of anesthesia. The rate may run from 30 to 75 per minute and the pulmonary ventilation from 60 to 150 cc.

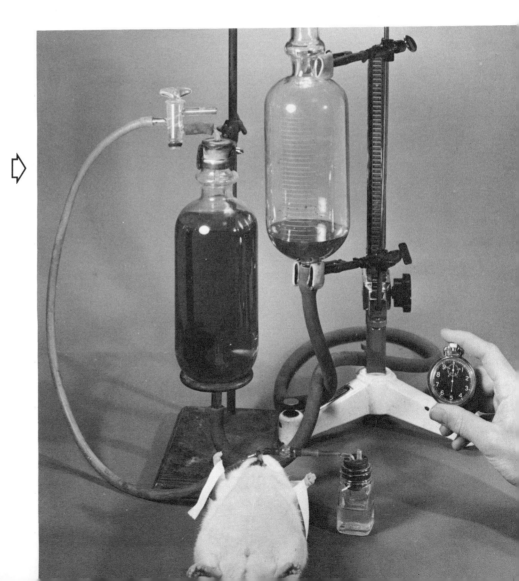

13 continued

STUDENT'S REPORT

EXPERIMENT 14

ACTION OF CARBON DIOXIDE ON RESPIRATION

In this experiment the stimulating action of carbon dioxide on respiration is shown by having the rat rebreathe expired air from a spirometer. That the increased respiration is due to carbon dioxide excess and not to oxygen lack is demonstrated by repeating the experiment under conditions in which the carbon dioxide is removed.

MATERIALS AND EQUIPMENT

Ink-writing lever and kymograph.
5 per cent sodium hydroxide.
Spirometer.

The wick, used for carbon dioxide absorption, is not shown in the diagram; it is described in the text.

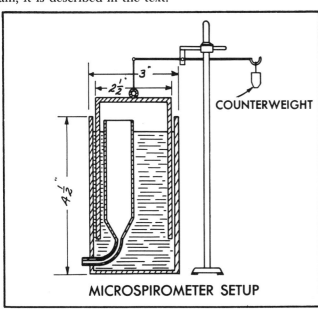

MICROSPIROMETER SETUP

PROCEDURE

1 Anesthetize the rat and cannulate the trachea.

2 Set up the spirometer and attach the tracheal cannula to the inlet. A wick, made by wrapping a piece of cloth around a small wire frame, is placed in the test tube into which the inlet opens and is moistened with weakly acidulated water. Start the drum. As the animal breathes, the concentration of carbon dioxide in the air within the spirometer will be increased, which results in an increase in respiration, as regards both rate and depth. If desired, the record may be timed with the timer so that the increased rate may be quantitatively determined

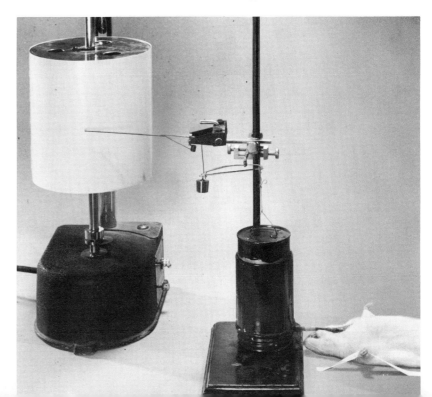

14 continued

and the throw of the needle be calibrated by means of a syringe attached to the inlet.

3 The increase recorded might have been due either to the carbon dioxide excess or to oxygen deficiency. To prove that it is the former, disconnect the rat from the spirometer, remove the bell, which permits the accumulated carbon dioxide to escape, and moisten the wick with 5 per cent sodium hydroxide solution. Again attach the rat and obtain another record. Little or no effect on respiration is now observed, during the same time interval as before, since the carbon dioxide is now absorbed by the sodium hydroxide.

RESULTS—SAMPLE RECORDS

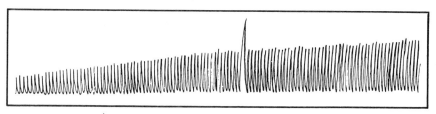

Record showing increase in depth of respiration from rebreathing air.

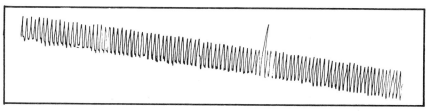

Animal is rebreathing air, but the carbon dioxide is being absorbed by sodium hydroxide; the respiration is not altered. The fall in the general level of the graph is due to the decreased volume of air in the spirometer as carbon dioxide is removed.

STUDENT'S REPORT

EXPERIMENT 15

THE HERING-BREUER REFLEX

The depth of respiration is controlled, in part, by nerve impulses passing up the vagi from the lungs. The impulses check inspiration when a certain degree of inflation has been reached; by cutting the vagi, these impulses are interrupted, and respiration becomes deeper and slower.

MATERIALS AND EQUIPMENT
Pneumograph and tambour.
Usual operating instruments.

PROCEDURE

1 Anesthetize a rat and place loops of thread around the vagi. Handle gently.

2 Fasten the pneumograph around the rat's thorax and connect to the tambour. Take a normal tracing.

3 While the record is being taken, lift one vagus by the loop and cut. Note the effect.

4 Then do the same for the other vagus. A pronounced deepening and slowing of respiration will be observed.

The experiment may also be performed by fastening a bent pin to the xiphoid process and attaching a thread from the pin to a writing lever; or the spirometer used in Part I, Experiment 14, with the wick saturated with sodium hydroxide, may be used.

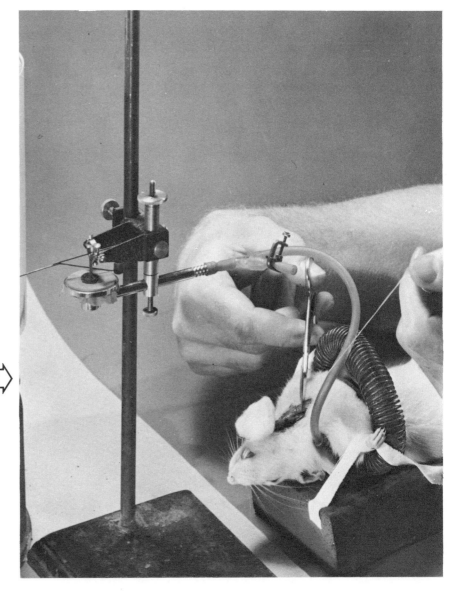

continued

RESULTS—SAMPLE RECORDS

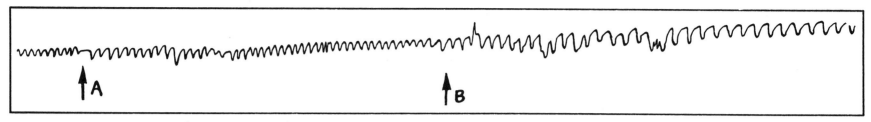

Record of respiration from pneumograph. At *A* one vagus was severed; at *B*, the other. Note slowing and deepening of respiration.

STUDENT'S REPORT

EXPERIMENT 16

EFFECT OF STIMULATING THE PHRENIC NERVE

The phrenic nerves are motor nerves which carry impulses to the diaphragm. In this experiment this action is demonstrated by applying a stimulus to the nerve near its origin in the neck. The resulting contraction of the diaphragm is recorded by means of an attachment connecting the diaphragm and writing lever.

MATERIALS AND EQUIPMENT

Ink-writing lever.
Diaphragm attachment.
Inductorium.
Dissecting instruments.

PROCEDURE

1. To locate the left phrenic nerve, make a mid-line incision in the neck as for carotid exposure, but retract the skin farther than usual to the left side. Dissect deeply to the origins of the cervical spinal nerves. It is convenient to free the sternomastoid from its lower attachment.

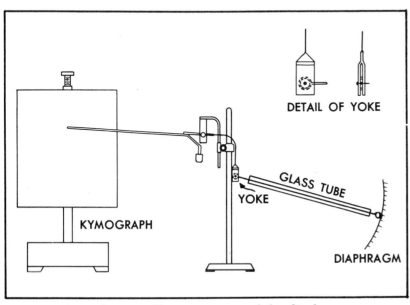

Setup for recording movements of the diaphragm.

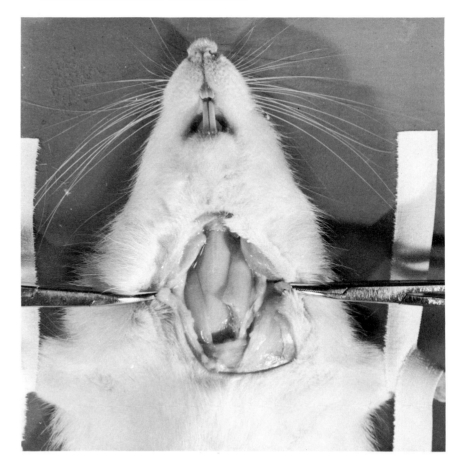

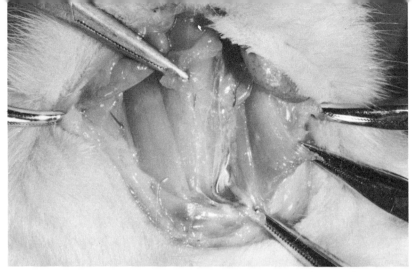
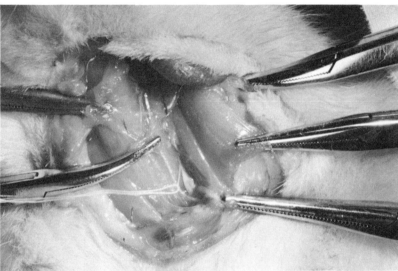
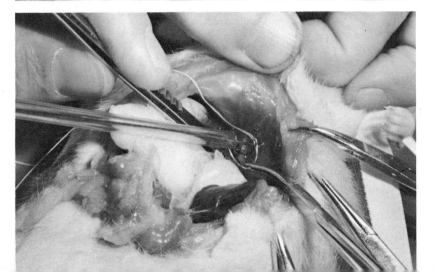

16 continued

2 The phrenic arises from the fourth and fifth cervicals and descends into the thorax, crossing over the fan-shaped seventh and eighth almost beneath the clavicle. This portion of the nerve is most suitable for freeing and stimulation.

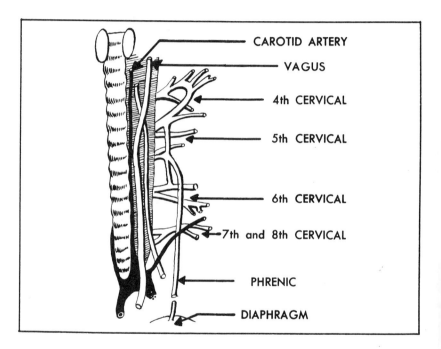

3 Place a loop of thread around the nerve and close the incision temporarily.

4 Make an abdominal incision slightly to the left of the mid-line. Gently push the viscera to one side. By means of a curved needle, pass a thread through the central portion of the left side of the diaphragm and fasten the thread tightly to the end of the wire of the diaphragm attachment.

5 Return the viscera to their place and close the abdominal incision, with the glass tube projecting upward at a slight angle. See that the wire moves freely in the glass tube with every excursion of the diaphragm. Bring the rat to the proper position in respect to the writing lever and slip a pin through the yoke. Obtain a normal record.

6 Return now to the neck, open the incision, free as much of the phrenic nerve as possible, and tie as far headward as possible. Cut headward to the tie.

7 Stimulate the peripheral end with short bursts of tetanizing current and also with individual induced shocks. Powerful contractions of the diaphragm will be recorded.

Cutting the phrenic nerve as described in this experiment does not paralyze the diaphragm because of the bilateral distribution of the nerve fibers.

RESULTS—SAMPLE RECORDS

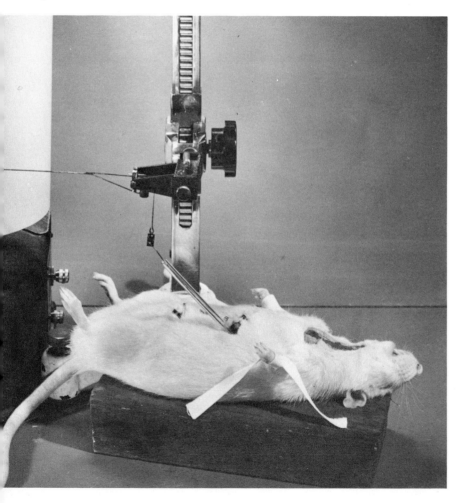

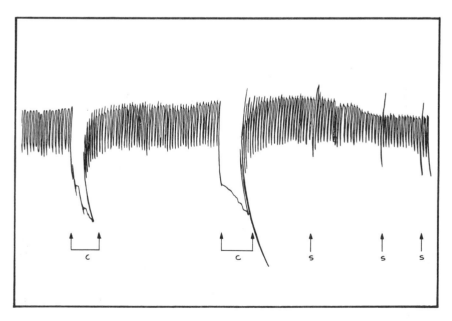

Record of phrenic nerve stimulation, with continuous stimulation between arrows at *C*; single shocks at *S*.

16 continued

STUDENT'S REPORT

EXPERIMENT 17

EFFECTS OF VAGAL STIMULATION ON GASTRIC AND INTESTINAL MOTILITY

MATERIALS AND EQUIPMENT

Inductorium. (See also Part I, Exper. 34.)

Kymograph.

Tambour and ink-writing lever.
> Pressures will be measured in centimeters of water; therefore membrane must be sensitive.

1 per cent barium chloride solution.

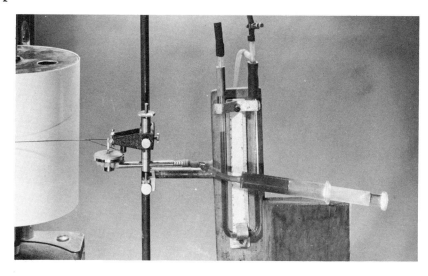

PROCEDURE

Set up tambour and kymograph; calibrate the drum up to 6 cm. of water, at intervals of 1 cm.

Stimulation of gastric motility

1. Anesthetize a rat, make a mid-line abdominal incision extending tailward about 2 cm. from the xiphoid process.

2. Extend the incision laterally in both directions from the upper end for a short distance. The viscera must be gently handled to avoid shock; be generous with the use of warm saline packs.

3. Dissect carefully to expose the cardiac sphincter and then follow the esophagus upward toward its emergence through the diaphragm. The liver can be pushed aside with cotton packs and retractors.

4. Free the two vagi, found on either side of the esophagus, and place a loop around the left vagus. (The experiment can be done by using the vagi in the neck, but cardiac effects are avoided by the present method.)

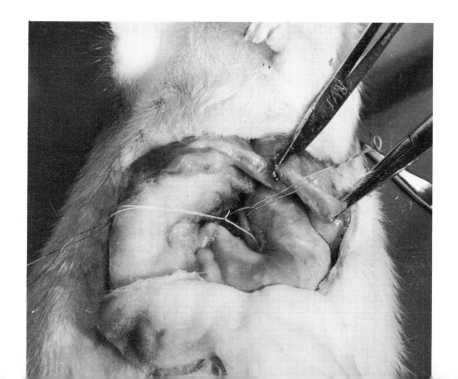

continued

5 Tie off the esophagus tightly, being careful not to include the vagi.

6 The upper part of the incision is now closed temporarily with wound clips.

7 Extend the lower part of the incision somewhat and expose the pyloric sphincter.

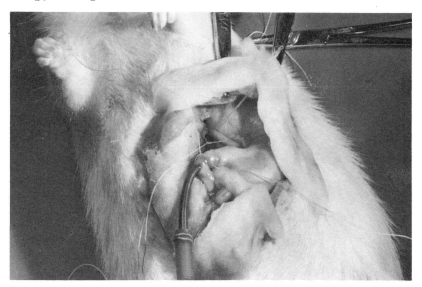

8 Make a snip in the duodenum and introduce a bent cannula through the sphincter, being sure that the tip is free within the stomach. Tie tightly.

9 Alternately fill and empty the stomach through the cannula to remove the gastric content, then fill with saline solution by means of a syringe. All saline used must be warmed to body temperature.

10 Attach the cannula to the tambour and, by means of a T-tube, fill the stomach with saline until slightly distended; usually a pressure of 4–6 cm. of water will give a record of good normal tonus waves.

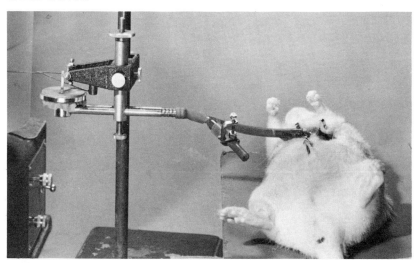

11 Close the lower part of the incision with wound clips but do not disturb the position of the cannula.

12 Return to the vagus. Tie the loop as far headward as possible and cut headward to the tie. Using the thread to hold up the nerve, stimulate the distal end while a normal record is being taken, with a moderately strong stimulus, continued for 10 seconds.

The response to vagal stimulation varies with the tonus present and the strength of stimulus. The usual response is a pronounced relaxation of the stomach at the moment of stimulation, followed by tonus waves of greatly increased strength.

Stimulation of intestinal motility

If the rat is in good condition, it may be used for the second part of the experiment; it is better, however, to use another rat.

PROCEDURE

1 Tie off the duodenum at the pyloric sphincter and gently free a loop about 8 cm. long.

2 Cannulate at the lower end and fill with warm saline.

3 Attach the cannula to the tambour and take a normal tracing, varying the pressure until a good record is shown. Continue the record for several minutes and note the rhythmic change in strength of the tonus waves.

4 Having prepared the vagus as above described, stimulate it. The effect is not so pronounced as on the stomach and also varies with conditions. Usually a slight fall is followed by a series of strong waves, the series being of considerably longer duration than is the normal rhythm.

Stimulation of intestinal loop

It is interesting to provoke and observe the peristaltic waves produced by direct stimulation of the intestinal loop. Little change in pressure is produced by these peristalses. To terminate the experiment, soak a cotton pad with 1 per cent barium chloride solution and place on the cannulated loop. Intense contraction of the viscus follows, producing a great increase in pressure.

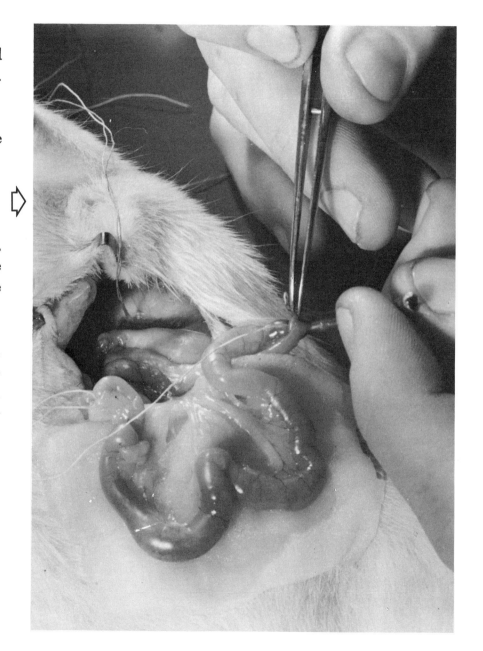

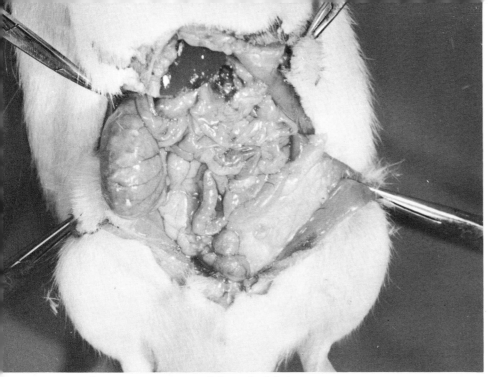

Normal intestinal loop.

Contracted intestinal loop after application of 1 per cent barium chloride solution.

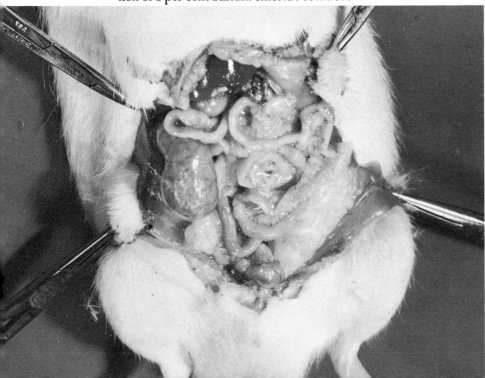

 **continued**

RESULTS—SAMPLE RECORDS

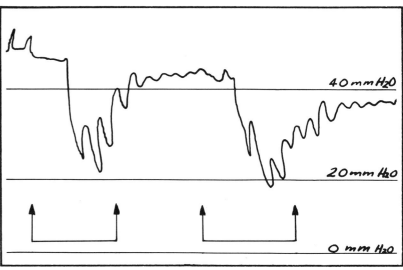

Effect of vagal stimulation on the stomach. Note initial relaxation, followed by strong tonus waves.

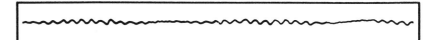

Normal rhythm of tonus waves in duodenum.

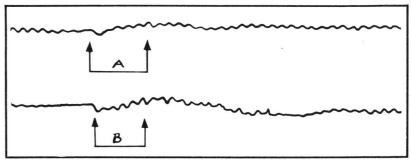

Effects of vagal stimulation. Note increased duration of active period (A), and increased strength of waves (B).

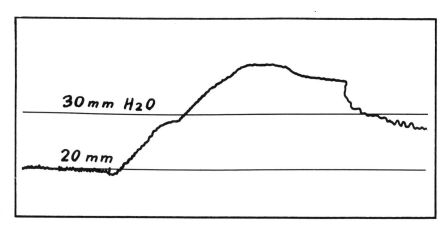

Effect of placing 1 per cent barium chloride solution on loop of duodenum.

STUDENT'S REPORT

EXPERIMENT 18
EFFECTS OF VAGAL STIMULATION AND OF HISTAMINE ON GASTRIC SECRETION

MATERIALS AND EQUIPMENT

Inductorium.

Histamine solution; 100 mg/cc.

Standard sodium hydroxide, approx. 0.01 N.

Phenolphthalein, 1 per cent in 50 per cent alcohol.

Preparation of standard sodium hydroxide, approximately 0.01 normal:

Dissolve 0.4–0.5 gm. of NaOH in enough water to make 1 liter of solution, mix thoroughly.

Standardize the solution by weighing accurately on an analytical balance, 0.100 gm. of potassium acid phthalate, transfer to a 250-cc. Erlenmeyer flask, and dissolve in about 50 cc. of distilled water. Add 3–4 drops of phenolphthalein indicator and titrate with the NaOH solution until a pink color which lasts 30 seconds is obtained. Record the volume of NaOH solution required. Repeat the standardization until check results are obtained.

Calculations:

$$\text{Normality of NaOH} = \frac{0.100}{0.204 \times \text{volume of NaOH used}}$$

PROCEDURE

Two rats are used, Rat I for vagal stimulation and Rat II for histamine injection.

1 Anesthetize the rats.

2 In Rat I, expose one of the vagi in the neck.

3 Cannulate the pyloric end of the stomachs of Rats I and II as described in the previous experiment and, by means of a syringe filled with saline warmed to body temperature, thoroughly wash out the gastric content.

4 Clamp off the cannulae and close the incisions with wound clips.

5 Wait exactly 30 minutes, keeping the rats warm and the wounds covered with warm saline packs. Then wash out the stomachs thoroughly and titrate the washings with standard NaOH to determine the normal secretion of hydrochloric acid.

6 Clamp off the cannulae and close the wounds as before.

7 In Rat I, tie off and cut the vagus.

8 During the ensuing 30 minutes, stimulate the distal end for 5-second periods, at 20-second intervals.

9 At the end of that time, again wash out the stomach and titrate the washings.

10 Inject Rat II subcutaneously with 100 mg/kg of histamine.

11 Wait 30 minutes, wash out stomach, and titrate the washings.

RESULTS OF TYPICAL EXPERIMENT

The results of a typical experiment are as follows:

Normal secretion, 30 minutes: Equiv. of 0.4 cc. 0.01 N sodium hydroxide.

Vagal secretion, 30 minutes: Equiv. of 3.1 cc. 0.01 N sodium hydroxide.

Histamine secretion, 30 minutes: Equiv. of 1.8 cc. 0.01 N sodium hydroxide.

STUDENT'S REPORT

EXPERIMENT 19

THE SECRETION AND DIGESTIVE ACTION OF SALIVA

In this experiment the fact is demonstrated that saliva from the parotid has a higher concentration of amylopsin than that from the submaxillary gland. It is also shown that the sympathetics to the submaxillary are not secretory in their action.

MATERIALS AND EQUIPMENT

Dissecting microscope.

Submaxillary duct cannulae are made from P.E. 10 tubing (see Experiment 2).

Inductorium.

Spot plate.

Eye dropper.

Stock starch solution (1 per cent).

Grind 2 gm. of potato starch in a mortar containing 15 cc. of cold water. Stir until a uniform paste is obtained. Slowly pour the suspension into 185 cc. of boiling water, stirring meanwhile. The stock solution should be further diluted, 1 part solution to 9 parts water, for use in the experiment.

Iodine–potassium iodide solution.

Prepare a 2 per cent solution of potassium iodide and add sufficient iodine to color it a deep yellow.

Pilocarpine solution of 0.5 mg/cc.

PROCEDURE

Amylolytic Power of Submaxillary versus Parotid Saliva

1 Anesthetize the animal with Nembutal.

2 Make a mid-line incision in the neck and retract the skin, exposing the submaxillary glands. These are large, pale, multilobed structures located on either side of the trachea. The major sublinguals are pink glands lying on the latero-anterior

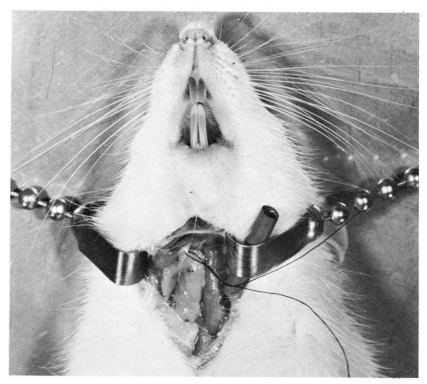

surface of the submaxillaries, which appear at first to be part of the submaxillaries. By carefully pulling apart the overlying sheath, the two glands may be separated from each other and their respective ducts clearly seen.

3 The following step is best carried out under the dissecting microscope. Tie off both sublingual ducts (ducts of Rivinus) and one submaxillary duct (Wharton's duct). Free the remaining submaxillary duct as far headward as possible. Place a tie around the head end, and, while exerting traction on the tie and on the caudal end of the gland, cannulate the duct. In order to make certain that the cannula is in the duct and not in the sheath, attach a syringe to the cannula and force a little saline in and out of the duct. If the cannula is properly inserted, the duct and its branches will swell for a considerable distance into the gland. After this test, the saline should be removed completely from the cannula.

4 Cannulate the opposite jugular and, after wiping the rat's mouth free of saliva, inject 1.0 mg/kg of pilocarpine. This will be followed by a copious flow of saliva.

5 Collect the saliva from the submaxillary in a small test tube. With the aid of a syringe, withdraw the mouth saliva (which will represent parotid secretion, as both submaxillaries and the major sublinguals are tied off), and place it in another test tube. Dilute the parotid saliva with saline until it is thin enough to be handled adequately with an eye dropper without "stringing." (It will be necessary to dilute it at least five times.)

6 Add 1 drop of submaxillary saliva to 4 drops of starch and test with the iodine on a spot plate at intervals of 15, 30, 60, and 120 seconds. Digestion is measured by a transition of colors from blue (indicating no digestion) to colorless (indicating complete digestion). Repeat the procedure with the parotid saliva and compare results.

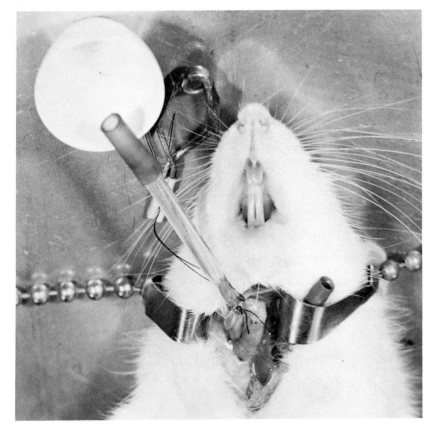

19 continued

RESULTS OF TYPICAL EXPERIMENT

Submaxillary saliva, stimulated by pilocarpine: No reaction in 2 minutes.

Parotid saliva, stimulated by pilocarpine: Reaction complete in 15 seconds.

EFFECT OF SYMPATHETIC STIMULATION

Having noted the increased flow of saliva from the submaxillary gland following the injection of pilocarpine (a parasympathetico-mimetic drug) in the above experiment, it may be interesting to show that the sympathetics do not increase the salivary flow, probably being trophic in their action.

PROCEDURE

Anesthetize another rat with Nembutal.

Cannulate the right submaxillary duct.

Free the right cervical sympathetic nerve. It is most easily located in the region of the bifurcation of the carotid, where the large superior cervical ganglion lies immediately between the internal and the external carotid.

Stimulate the sympathetic with long-continued shocks.

RESULTS OF TYPICAL EXPERIMENT

No, or very little, increase in the rate of salivary secretion will be noted.

STUDENT'S REPORT

EXPERIMENT 20

THE SECRETION AND DIGESTIVE ACTION OF PANCREATIC JUICE

In the first part of this experiment, the action of secretin in stimulating pancreatic juice secretion is demonstrated. In the second part, the digestive action of trypsin on protein is studied.

MATERIALS AND EQUIPMENT

Intestinal cannulae.

Bile duct cannula (PE-10 polyethylene tubing plus 32-gauge wire).

Mett's tubes.

Capillary tubes such as used for blood coagulation, in Part I, Experiment 6, may be used.

Incubator (at 37°–40° C.).

50 cc. of 5 per cent hydrochloric acid.

50 cc. of 10 per cent sodium hydroxide.

White of egg.

Pilocarpine solution, 0.75 mg/cc.

Buffer solution, pH 8.4.

Dissolve 12.405 gm. of boric acid and 14.912 gm. of potassium chloride in 1 liter of water. To 50 cc. of this solution add 8.5 cc. of 0.2 N sodium hydroxide.

PROCEDURE. Part I

1 In the rat, many of the pancreatic ducts empty into the bile duct. This may be demonstrated by cannulating the bile duct as described below and injecting a solution of methylene blue, which, by filling the pancreatic ducts, makes them readily visible. This should have been done previously on a dead rat, as a demonstration.

2 A rat is anesthetized with Nembutal, and one femoral vein cannulated. A transverse incision is made, beginning at the border of the lower ribs on the right side and extending to the

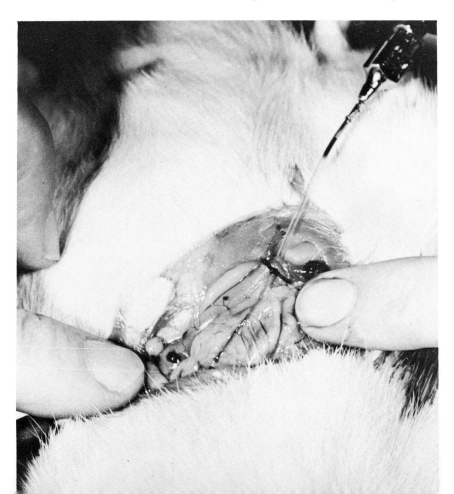

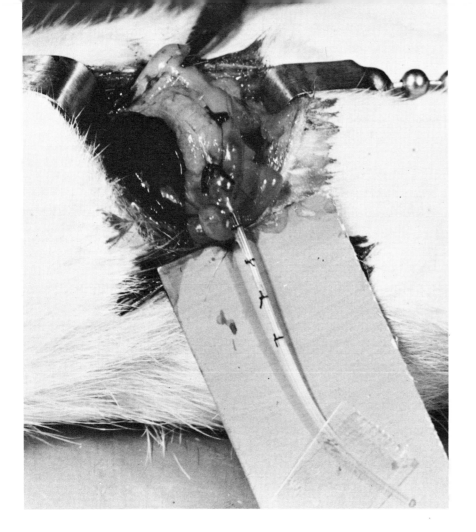

20 continued

made very low, in order to prevent tying off one of the largest of the entering pancreatic ducts.)

3 The pancreatic juice rises in the cannula, and the rate of flow is determined by timing its rise past 2 or 3 of the marked divisions of the cannula. This represents the normal secretion rate.

4 Preparation of secretin. Another rat is sacrificed, and the entire small intestine is excised. Fill the intestine with 5 per cent hydrochloric acid and allow to stand for 1 hour. Remove the intestinal content, boil for several minutes, neutralize with 10 per cent sodium hydroxide (litmus), and filter.

5 The secretin solution is warmed to body temperature, and 1 cc. is slowly injected into the jugular vein.

RESULTS OF TYPICAL EXPERIMENT

The usual rate of pancreatic secretion, using the type of cannula described, is 5 mm. every 10–15 minutes. Within 10 minutes following the injection of secretin, the rate will be increased two- to fivefold, and the increased rate maintained for from 15 to 20 minutes.

PROCEDURE. Part II

1 Another rat is anesthetized, and a transverse incision made as described above. A loop of the small intestine is now cannulated, one cannula being inserted just below the pyloric sphincter and the other about 2 cm. below the entrance of the bile duct.

mid-line. Locate the bile duct and tie off its upper end just below its bifurcation. Trace the duct toward the intestine. The lower end can be seen lying along the surface of the intestine. Now make a cut in the intestinal wall just below the point where the bile duct disappears. Cannulation is performed by inserting the tip of the cannula into the bile duct through the cut in the intestinal wall. A thread is passed around the bile duct by means of a curved needle, and the cannula tied in place. (It is necessary that the cannulation be

2 The upper end of the bile duct is tied off just below its bifurcation, to prevent the entrance of bile.

3 The intestinal loop between the cannulae is washed out with warm saline, which is discarded.

4 The intestinal cannulae are now clamped off for 30 minutes, following which the intestinal loop is washed out with exactly 5 cc. of warm saline. This represents the control enzyme solution.

5 Now introduce directly into the abdominal cavity 1 cc/kg of the pilocarpine solution, wait 30 minutes, and again wash out the intestinal loop with 5 cc. of warm saline. This represents the experimental solution.

6 Strain the white of an egg through fine cheesecloth and draw it up into the capillary tubes. Hold the tubes horizontally and drop them into water at 85° C. Let the water cool. Seal the end of the tubes with paraffin, in which condition they will keep indefinitely.

7 To the control and experimental enzyme solutions now add equal volumes of the buffer solution. Cut the capillary tubing containing the coagulated egg white into 2-cm. sections and place them in small test tubes which have been filled with the enzyme solutions. Place them in the incubator (37°–40° C.) for 48 hours.

RESULTS OF TYPICAL EXPERIMENT

Examination of the Mett's tubes at the end of 48 hours will show little, if any, digestion resulting from the action of the control solution, while the tubes incubated in the experimental solution will show from 2 to 4 mm. digestion.

STUDENT'S REPORT

EXPERIMENT 21

EFFECT OF DECHOLIN ON BILE FLOW

In this experiment the effect of a choleretic, dehydrocholic acid (Decholin), on the flow and concentration of bile is shown.

MATERIALS AND EQUIPMENT

Cannula. As in previous experiment.
Stop watch.
Decholin tablets (0.25 gm.).

PROCEDURE

1. Make an abdominal mid-line incision extending from somewhat below the xiphoid process about 3 cm. tailward.

2. Expose the duodenum and locate the bile duct, a small duct extending from the median lobe of the liver and emptying into the duodenum a short distance below the pyloric sphincter. Free a portion of the duct and cannulate toward the liver.

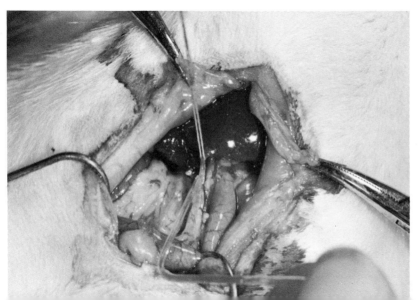

3. Close the incision with wound clips and turn the animal on one side so that the cannula extends out of the body and downward. Make sure that the duct and cannula are straight in line. Count the number of drops, over 5-minute periods, for three such periods, or until a constant figure is obtained. Save the bile.

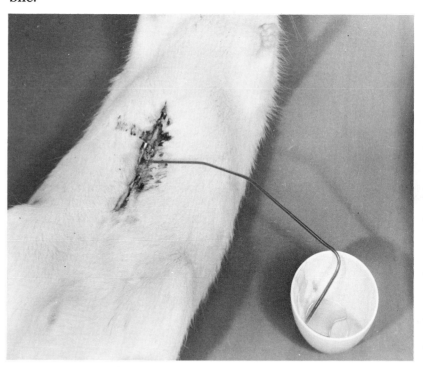

4. Dissolve a Decholin tablet in 2 cc. of warm saline and inject with a large needle into the duodenum anterior to the bile duct. Count the drops for three 5-minute periods. Save the bile secreted during the period of maximum flow.

5 Determine the density of the two samples, using the same method as was described for blood (Part I, Exper. 6). It will be found that the action of this choleretic is to increase markedly the volume of bile secreted but that this bile is less concentrated than the normal bile.

RESULTS OF TYPICAL EXPERIMENT

Normal flow: 4 drops/5-min. interval.
Stimulated flow: 9 drops/5-min. interval.
   (The absolute number of drops will, of course, depend on the bore of the cannula used.)
Density of normal bile: 1.04–1.05.
Density of stimulated bile: 1.024–1.04.

STUDENT'S REPORT

EXPERIMENT 22

SALINE DIURESIS

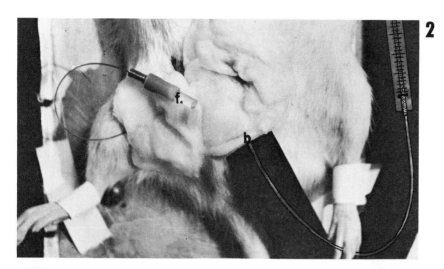

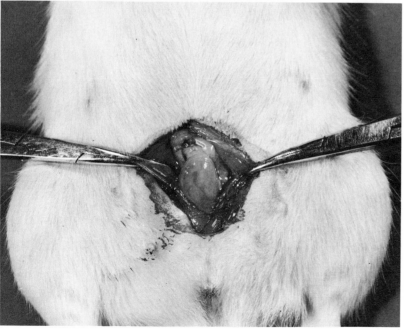

In this experiment the effect of intravenous injections of saline and of a 10 per cent glucose solution upon the rate of urinary secretion is observed.

MATERIALS AND EQUIPMENT

PE-50 polyethylene cannula with 26-gauge stiffening wire.

Rate meter. This is made by fastening a piece of capillary tubing to a centimeter scale (**r**).

Female rat

In order to insure an adequate urine flow inject 1.5–2 ml. of 0.4 saline intraperitoneally several hours before the experiment is to be performed.

PROCEDURE

1. Make a mid-line incision through the lower third of the abdomen and expose the pelvic viscera. The bladder lies just anterior to the symphysis pubis.

2. Cannulate a femoral vein, heparinize (**f**), and close the wound temporarily with clips.

3. It is not necessary to cut the symphysis; by carefully clearing away the overlying muscle and connective tissue, a sufficient length of the urethra can be exposed. Cannulate, being sure that the tip of the cannula lies freely within the bladder, remove strengthening wire, and attach the other end of the cannula to the rate meter (**r**).

4 Close the wound with clips, being careful not to disturb the position of the cannula. Keep the wound warm and moist, but do not exert pressure on it.

5 Observe the urinary flow during a 15-minute period; there will be very little, if any.

6 Return to the femoral. Inject 2 cc. of saline solution, warmed to body temperature, slowly into the vein. Record the rise in the flow meter at 2-minute intervals during the next 16 minutes. If the first 2 cc. of saline are not effective, repeat the injection.

7 After the diuresis has ceased and the flow has returned to normal, inject 1 cc. of a 10 per cent glucose solution. (The student should realize that this will increase the rat's blood sugar from the normal of approximately 100 mg. per 100 cc. of blood to 600 mg.) An even more marked diuresis will follow. (The rat is highly resistant to the diuretic action of caffeine and relatively resistant to urea.)

RESULTS OF TYPICAL EXPERIMENT

	Drops
15-minute period before injection	None
First 5-minute period after saline injection	2
Second 5-minute period after saline injection	5
Third 5-minute period after saline injection	5
Fourth 5-minute period after saline injection	2
First 5-minute period after glucose injection	7
Second 5-minute period after glucose injection	5
Third 5-minute period after glucose injection	2

(The absolute number of drops depends, of course, on the bore of the cannula used.)

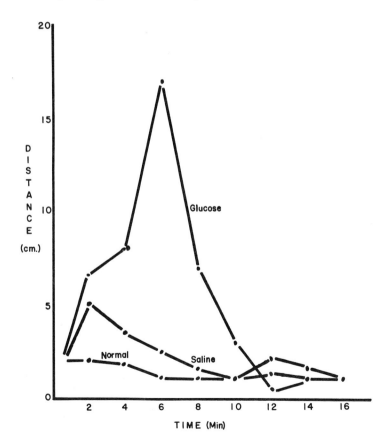

22 continued

STUDENT'S REPORT

EXPERIMENT 23

REFLEX CONTRACTION OF THE BLADDER

MATERIALS AND EQUIPMENT

Polyethylene PE-50 cannula plus 26-gauge wire.

Sensitive tambour.

Ink-writing setup.

PROCEDURE

1. Through a low, mid-line abdominal incision, expose the urinary bladder. Holding it up by the tip, make a tiny snip, with fine-pointed scissors, in the dorsal wall just under the tip and insert the cannula. Tie, but do not include any more of the bladder than is necessary in the tie.

2. Close the abdominal incision with wound clips.

3. Fill the cannula with saline and attach it to one arm of a T-tube whose other arm is attached to the tambour. Be sure the system is free of air bubbles.

4. To the remaining arm of the T-tube attach a 5-cc. syringe full of saline which has been warmed to body temperature. With the kymograph running at its slowest speed (the kymograph has been previously calibrated, at 5 cm. of water-pressure intervals), inject saline in amounts of 0.25 cc., at intervals of 1 minute. When between 1 and 2 cc. of saline have been introduced, or when the pressure is from 15 to 25 cm. of water, a distinct relaxation of the bladder occurs, with a consequent

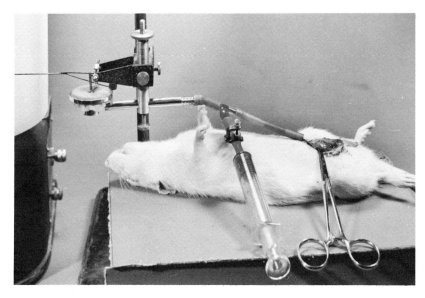

fall in pressure (adaptation reflex). Tonus waves will be noted during these intervals.

5. Further introduction of saline will produce the micturition reflex. The rate at which urine is voided depends mainly upon the degree to which the bladder has been distended, although when it occurs with small amounts of saline in the bladder there may be no great change in pressure. Careful observation will show a slight rise in pressure just preceding the expulsion of each drop of urine.

6. Acceleration of voiding will frequently occur following stimulation—by pinching or by cold—to the region of the anus or the vagina.

23 continued

RESULTS—SAMPLE RECORDS

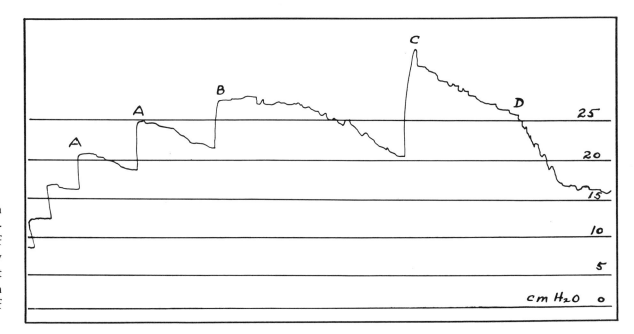

Record of bladder reflexes. Note adaptation reflex following points marked A. At B an increase in pressure followed the introduction of 0.5 cc. of saline, which was then followed by micturition. One cc. of saline was introduced at C, again followed by micturition; at D the skin around the anus was pinched, and the rate of voiding increased.

STUDENT'S REPORT

EXPERIMENT 24

THE EFFECTS OF ADRENALECTOMY AND THE ACTION OF CORTIN

The group of experiments on the endocrine system includes a number of "chronic" experiments, i.e., operations following which the animals are to be kept alive for an indefinite period. Such operations must be performed under sterile conditions, to avoid wound infection.

General directions for aseptic operations

Sterilization of pack. The required towels, thread, and cotton are wrapped in a piece of canvas. If no autoclave is available, an ordinary pressure cooker may be used. The pack should be placed in the cooker on a tray above the level of the water, and a steam pressure of 20 pounds maintained for 30 minutes. The escape valve is opened, and the bundle is removed while still hot. It will be somewhat moist but can be left overnight to dry.

All instruments used during the operation are kept in an evaporating dish containing 70 per cent alcohol. A similar dish of boiled saline is used to rinse the alcohol from the instruments before they are actually used in an open wound. Any instrument touching a nonsterile object is returned to the alcohol dish.

PREPARATION

In performing these operations, students work in pairs. While the surgeon scrubs his hands thoroughly, paying particular attention to the fingernails, the assistant anesthetizes the rat (since the operations are of short duration, ether alone

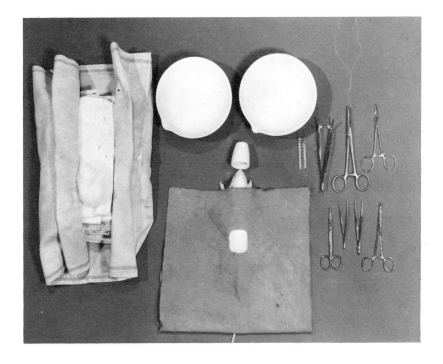

is used), clips the hair from the operation site, and scrubs the skin with soap and warm water and finally with alcohol. He then places the animal on a sterile towel, which the surgeon has removed from the pack; the operative site is covered with another towel in which a small hole has been cut and through which the operation is performed. Instruments may then be laid on the towel and the operation is started, the assistant being responsible for the anesthesia.

After the operation the animals should be kept warm for several hours by being placed in warm, dry towels in a box, until they are fully conscious. Later they should be kept in

continued

freshly cleaned cages. While still anesthetized they may be numbered according to the system described in Experiment 1, in order to identify the animals operated on by different students.

The wound clips used to close skin incisions may be removed a week or 10 days after the operation; healing will be complete by that time.

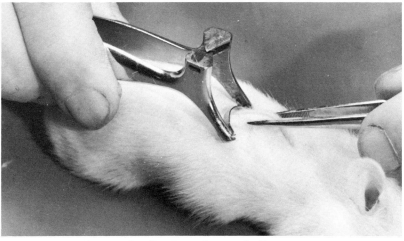

Application of wound clip.

Removal of wound clip.

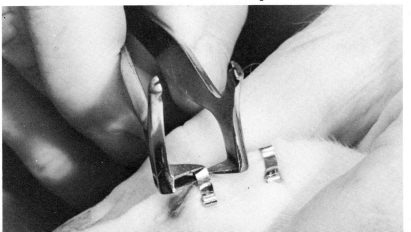

Experiment on the adrenal cortex

In this experiment the fact that the adrenal cortex is necessary to maintain life is demonstrated by removing the adrenal glands from a group of rats; half the group is left untreated and dies within a week, the other half receives a potent preparation of cortical hormone and is found to remain in good condition as long as the injections are continued. Each student should adrenalectomize at least two animals. Male rats weighing 45–55 gm. should be used.

MATERIALS AND EQUIPMENT

Sterile pack.

Surgical instruments.

Cortin preparation.

> Percorten acetate, a synthetic desoxycorticosterone acetate preparation produced by Ciba, is satisfactory. It comes in ampoules containing 5 mg/cc.

PROCEDURE

1 Prepare for the operation as described above.

2 With the rat in the prone position make a dorsal mid-line incision about 2 cm. long, extending from about the tenth thoracic to the third lumbar vertebra. Both adrenals can be removed from this central skin incision.

3 Retract the skin laterally to the right and, with a pair of fine-pointed forceps, force a small hole through the thin muscle wall just anterior to the upper pole of the kidney. Pull the forceps open to enlarge the hole. The adrenal is imbedded in fat and connective tissue, which is loosely attached to the upper pole of the kidney.

4 Grasping the fat surrounding the adrenal with one forceps, dissect it loose from the kidney with the points of another. (Do not touch the gland itself; it is friable and easily crushed.) In rats of this size the blood vessels supplying the adrenal are small and need not be ligated. Lift the mass of fat out of the wound and examine it to be sure it contains the adrenal.

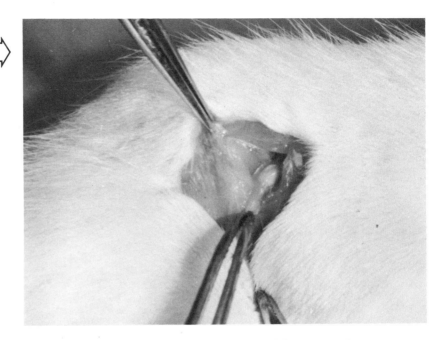

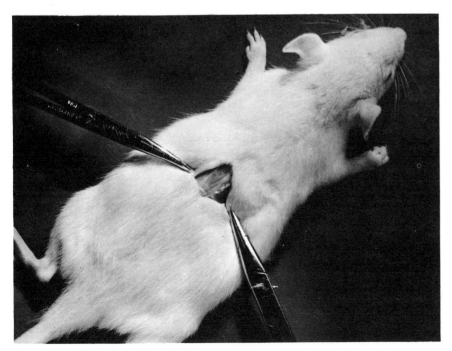

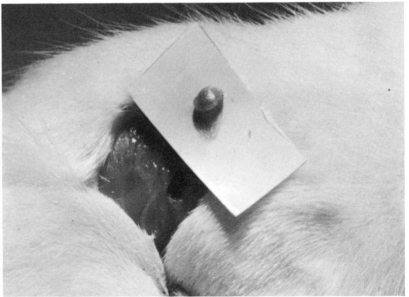

24 continued

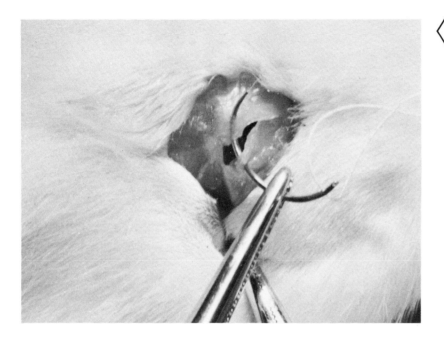

5 The muscle wound is closed with one or two stitches, using fine silk thread and a curved needle.

6 Repeat the operation on the left side and close the skin incision with wound clips.

7 The operated animals are divided into two groups; the untreated group is weighed daily and watched for symptoms of cortical deficiency. A progressive loss in body weight and, within a few days, anorexia, weakness, and a reduction in body temperature will be noted, ending in death, usually between the fifth and eighth days.

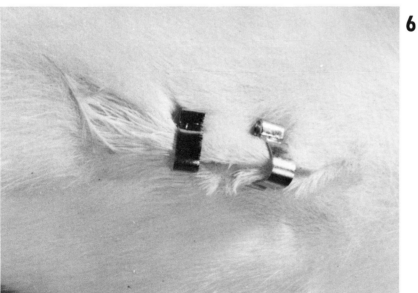

8 The members of the treated group are injected with 0.25 mg. of Percorten daily for a 10-day period, beginning the day after the operation. The injection is made subcutaneously, and the rats are weighed at the time of injection. This dose is adequate to maintain normal growth and health.

9 At the end of that period, injections are discontinued; the rats will deteriorate, and death will follow. An occasional rat will survive indefinitely, owing either to incomplete operative removal or to the existence of accessory cortical tissue.

STUDENT'S REPORT

EXPERIMENT 25

THE EFFECTS OF THYROIDECTOMY AND OF THYROXIN ADMINISTRATION ON THE BASAL METABOLIC RATE

In this experiment the normal B.M.R. of a group of rats is determined. They are then subjected to thyroidectomy, and, after an interval of several weeks, the B.M.R. determination is repeated. Thyroxin is then administered, and its ability to return the B.M.R. to (or above) the normal is demonstrated.

MATERIALS AND EQUIPMENT

Apparatus for determination of oxygen consumption.
Sterile pack and surgical instruments.
Thyroid preparation.

Thyroxin (Squibb), a preparation containing 2.0 mg. purified thyroxin per tablet, is satisfactory.

PREPARATION

Each student is assigned two adult rats; they should be placed in the cage in the desiccator, with the lid off, for several hours each day to become accustomed to the conditions. This saves time later on. Food should be withheld the evening before the actual determination is to be made.

PROCEDURE

To determine the oxygen consumption, the apparatus must first be calibrated.

1 The desiccator is filled with a 5 per cent sodium hydroxide solution (at room temperature) to just below the level of the cage in which the rat is to be placed, and a flask having about the same volume as the rat is placed in the cage.

2 With the tambour and writing lever adjusted, 20 cc. of air are forced into the desiccator, and a line is run around the drum.

3 Then 40 cc. of air are withdrawn, and another line is drawn.

4 The flask is now removed and filled with water at 40° C. and replaced in the cage. By means of a thermometer graduated in tenths of a degree, the volume equivalent of the rise in pressure, per degree rise in temperature, is determined. This correction is necessary in calculating the oxygen consumption.

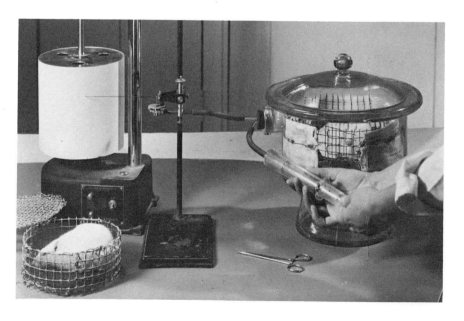

5 The rat is now placed in the chamber, 25–30 cc. of air are forced in, and the drum is started. As the carbon dioxide is absorbed, the line which is being written on the drum falls.

6 At the instant it crosses the upper calibration line, a stop watch is started and the temperature noted. When the lower calibration line is crossed, the time and temperature are again recorded. This provides the data for the calculation of oxygen consumption. If the rat moves about, the determination is repeated; in any case, at least three close checks on each animal should be obtained.

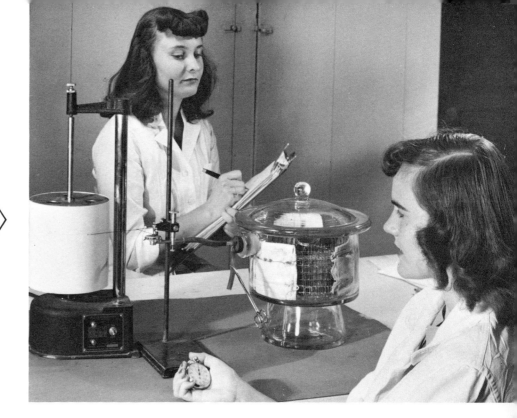

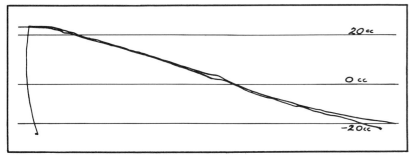

Duplicate records obtained on the same rat, showing a good check.

Thyroidectomy

1 Prepare for an aseptic operation as previously described.

2 Make a ventral mid-line incision through the skin of the neck, push the salivary glands to one side, and expose the trachea by slitting the overlying muscle in the mid-line. The thyroid is the reddish, bi-lobed structure lying on either side of the trachea just below the larynx. The parathyroids may be seen

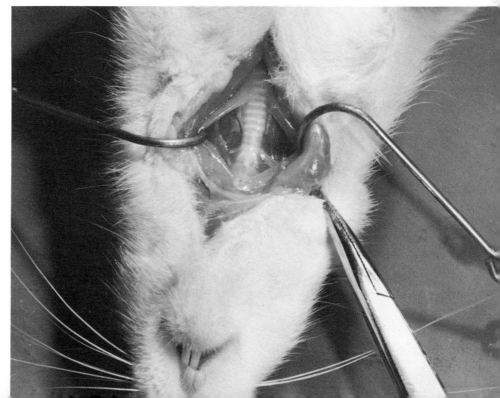

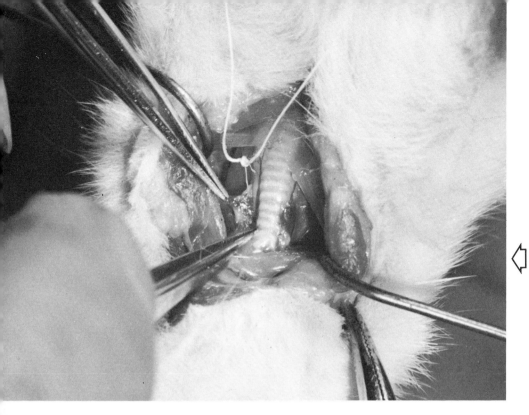
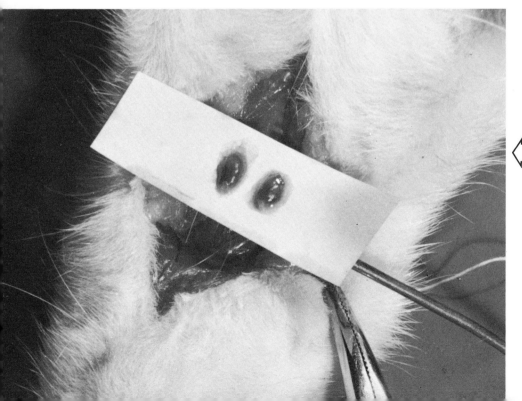

25 continued

as tiny, whitish bodies at the tip of the upper lobe. An effort should be made to leave them behind; in any case, however, calcium gluconate should be added to the drinking water for several days after the operation to avoid parathyroid tetany.

3 A small bridge of thyroid tissue (the isthmus) connects the two lobes across the trachea. With fine forceps divide the isthmus and, holding onto one end, dissect carefully underneath the lobe. Two structures must be watched for—the inferior thyroid artery, which is the main blood supply to the thyroid (the superior is smaller and may be disregarded), and the recurrent laryngeal nerve, which lies close to the trachea and passes below the lobe of the thyroid. The artery must be either ligated or clamped off with the tips of a fine hemostat; the nerve must not be disturbed at all, or respiratory difficulties will develop later, and the animal may die. With these precautions in mind, the lobe is dissected free, leaving a small amount at the upper tip inclosing the parathyroid, and removed.

4 The other lobe is then removed in the same manner.

5 Close the wound with fine silk sutures through the muscle and two or three more through the subcutaneous tissue.

6 Close the skin incision with wound clips. Watch the animals for a few days for signs of respiratory distress and give them a 1 per cent solution of calcium gluconate for drinking water.

7 After several weeks (it requires from 2 to 4 weeks for symptoms of hypothyroidism to become pronounced), the oxygen consumption is again determined as described above. A considerable fall (20–40 per cent) will have occurred.

Thyroidectomized animals are more sensitive to thyroxin than normal animals; therefore, these same rats are now used to demonstrate its action.

8 Crush a tablet of Thyroxyl in saline and inject, subcutaneously, a dose of 1 mg/kg. Redetermine the oxygen consumption 48–72 hours later. The oxygen consumption will be normal or somewhat above.

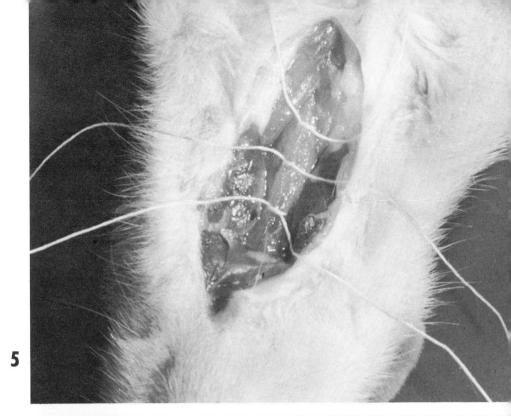

5

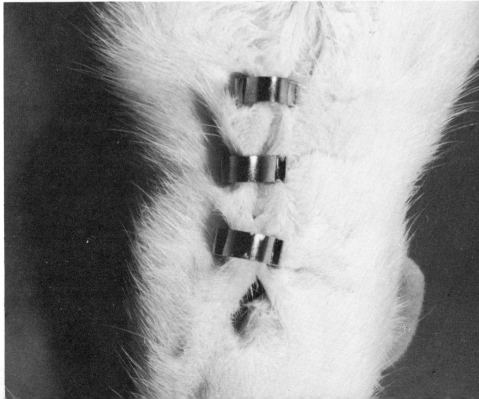

6

RESULTS OF TYPICAL EXPERIMENT

Normal rat; weight 200 gm. Consumed 40 cc. oxygen in 7 minutes 24 seconds. Temperature rise: $0.3°$. Room temperature: $28°$ C. (Volume equivalent per degree rise in temperature found to be: 15.0 cc.)

Total oxygen consumption: 44.5 cc.; oxygen consumption per minute, $\frac{44.5}{7.4} = 6.0$ cc.

Reduced to standard conditions: $6.0 \times \frac{630^*}{760} \times \frac{273}{301} = 4.54$ cc/min.

* Barometric pressure in Denver is 630 mm.

$4.5 \times 60 = 270$ cc/hr.
Assuming an R.Q. of 0.80, heat equivalent is 4.8 cal. per liter.
$4.8 \times 0.27 = 1.3$ cal/hr; surface area, $(200)^{2/3} \times 10 = 340$ sq. cm.

$\frac{1.3}{.034} = 38.2$ cal/sq m/hr.

		Calories
Average results:	normal	38–42
	thyroidectomized	30–32
	thyroidectomized plus Thyroxyl	40–44

25 continued

STUDENT'S REPORT

EXPERIMENT 26
THE EFFECTS OF OVARIECTOMY AND OF ESTRIN ADMINISTRATION ON THE ESTRUS CYCLE

In this experiment the endocrine function of the ovary is demonstrated by observing the effects of its removal upon the estrus cycle and upon the uterus. The ability of estrin (one of the ovarian hormones) to produce estrus effects in the ovariectomized animal is then shown.

MATERIALS AND EQUIPMENT

Microscope.

Sterile pack and surgical instruments.

Estrin preparation.
    Amniotin 10,000 I.U./cc (Squibb or Theelin 10,000 I.U./cc) (Parke-Davis). The dosage is given in International Units.

PROCEDURE

1 Each student is assigned 4 young adult (100–120-gm.) female rats. The estrus cycle is followed by observing the changes in types of cells in the vagina, the so-called "vaginal smear." To obtain this smear, tufts of cotton are wrapped tightly around the ends of toothpicks, the swabs moistened with saline and gently inserted and slightly rotated within the vagina. The swab is then pressed in a drop of saline on a microscope slide and examined under low power. From day to day the appearance of the cells in the smear is noted and recorded.

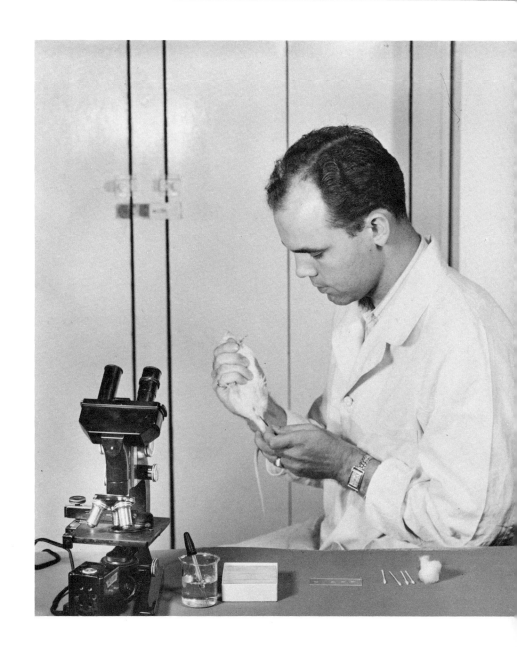

continued

Changes in vaginal smear during an estrus cycle

During the anestrus portion of the cycle the smear consists preponderantly of leucocytes, with an occasional cornified epithelial cell. The first appearance of estrus is marked by mucification of the vagina, followed quickly by the complete, or nearly complete, disappearance of leucocytes and their replacement by large numbers of round, nucleated, epithelial cells. These cells are about three times as large as the leucocytes and, once seen, cannot be mistaken. At this time and for about 12 hours afterward, the animals will breed, but at no other time. Desquamation of the epithelial cells now occurs, and white, cheesy masses of disintegrating squamous cells are found in the smear. Leucocytes now again make their appearance, and the cycle is repeated. The whole cycle in the rat lasts about 5 days, being divided roughly into 3 days of anestrus and 2 spent in the manifestations of estrus just described. The cycle makes its appearance at puberty, which is reached in the rat at an age of 60–90 days.

2 After the smears have been followed during one complete cycle, 2 of the 4 rats are sacrificed during the second cycle, one in estrus, the other in anestrus. The uteri are examined, and it will be seen that during anestrus the uterus is thin and pale, while in estrus it is pink, fleshy, and may be distended with fluid.

Having made and recorded the above observations on the normal estrus cycle, the remaining rats are ovariectomized.

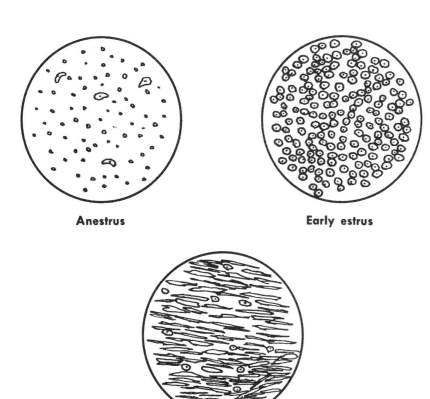

Anestrus

Early estrus

Late estrus

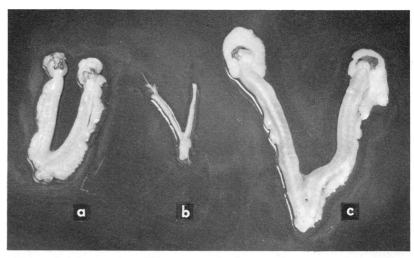

Uterus (*a*) in anestrus, (*b*) one month after ovariectomy, (*c*) in estrus.

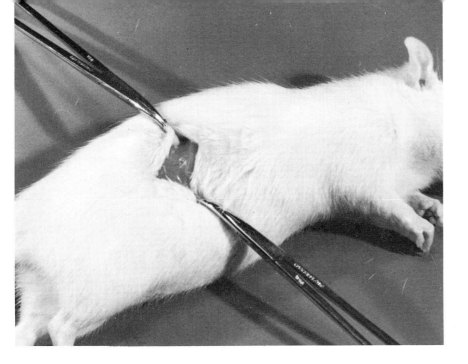
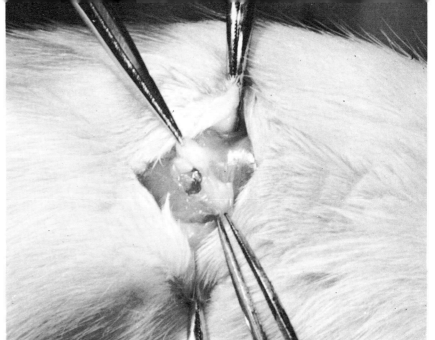

Ovariectomy

Prepare for an aseptic operation as described in Part I, Experiment 24.

1 With the rat lying on its side, make a skin incision running lengthwise, about 2 cm. lateral to the mid-line and running approximately from in front of the hip to a point somewhat anterior to the upper pole of the kidney. (It is better, at first, to make a fairly long incision, until the exact location of the ovary has been determined with practice.)

2 Make a snip through the fascia of the abdominal rectus muscle slightly lateral and anterior to the upper pole of the kidney, force the points of a forceps through the snip, and extend the hole by opening the forceps. The ovary will be found imbedded in fat lying just below the dorsal muscle mass. If the ovary is not immediately seen, the uterus probably will be; the ovary can then be located by following the uterus forward.

3 The ovary is drawn through the incision, the uterus clamped in a hemostat, and a ligature placed around the uterus just below the Fallopian tube and tied tightly. The uterus is then cut through on the ovarian side of the tie, and the ovary is removed.

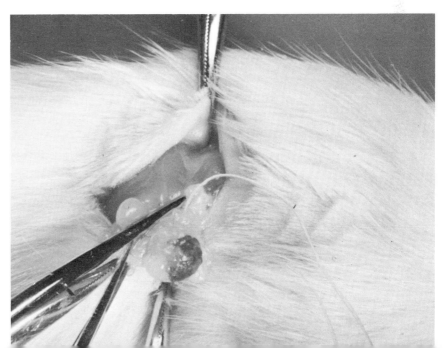

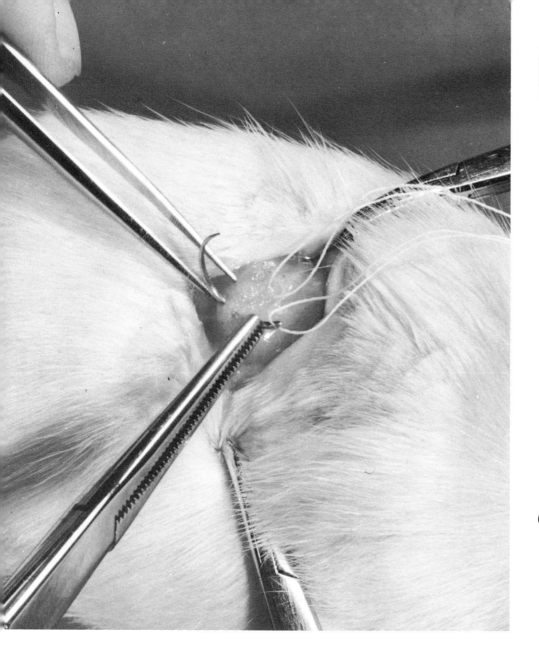

26 continued

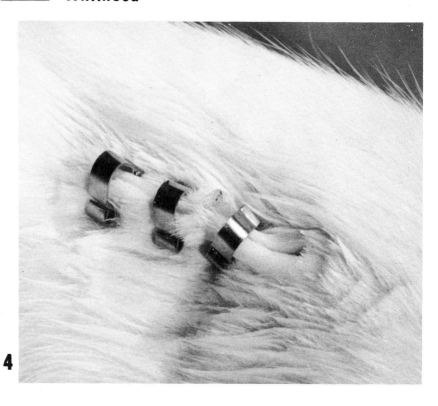

4

4 Close the muscle incision with the necessary number of silk-thread sutures and the skin incision with wound clips.

5 The other ovary is then similarly removed.

6 A week or 10 days is allowed for recovery from the operation, following which all animals are smeared daily for 1 week. It will be found that the estrus cycle has ceased. Half the rats are now injected with 200 I.U. of the estrin preparation. Vaginal smears are made 48 hours later; a smear typical of estrus will be obtained. These animals, as well as the uninjected, ovariectomized group, are now sacrificed, and the uteri inspected; those of the injected group will be large and distended, comparable to the normal uterus during estrus, while those of the uninjected group will be pale and thread-like in appearance.

STUDENT'S REPORT

EXPERIMENT 27

THE EFFECTS OF CASTRATION AND THE ACTION OF TESTOSTERONE

In this experiment one endocrine function of the testes is demonstrated by showing (1) that castration results in a failure of growth of the seminal vesicles and (2) that this growth is restored by testosterone, the testicular hormone.

MATERIALS AND EQUIPMENT

Sterile pack and surgical instruments.
Testosterone preparation.

    Oreton, a synthetic testosterone preparation made by Schering, containing 25 mg. per ampoule, is satisfactory.

PREPARATION

Three groups of young (75–100-gm.) male rats are used; one group serves as controls. The other group is castrated; half the castrates are untreated, the other half are injected subcutaneously with 0.1 mg. of testosterone daily for a 10-day period. At the end of this time, all rats are sacrificed and the seminal vesicles dissected out and weighed. Comparison of the weights will show that the seminal vesicles of the untreated, castrate group have decreased in size as compared with the normals but that testosterone was able to maintain, or even increase, the weight of the seminal vesicles in the castrate group which received it. The operative and injection work is divided equally among the students.

PROCEDURE

Castration

1. Prepare for an aseptic operation.

2. The rat is anesthetized and a ventral, mid-line incision made through the skin of the scrotum. Rats are able to retract the testes into the abdominal cavity; slight pressure over the pelvis will, however, force them back into the scrotum.

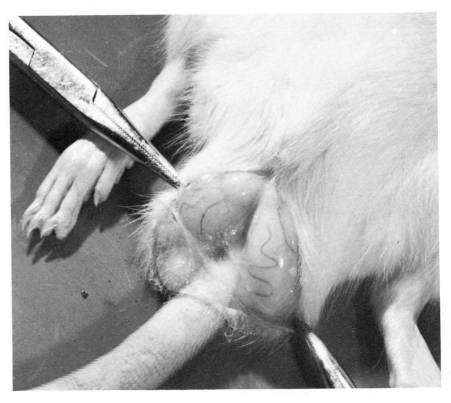

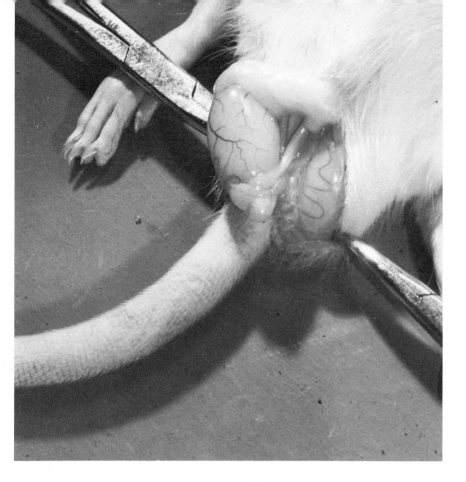

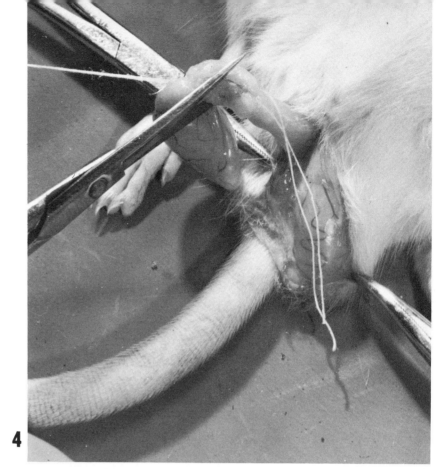

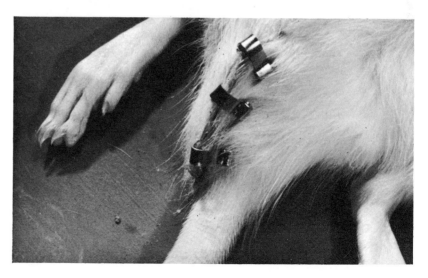

3 They are freely movable within the scrotum, and one testis is drawn through the skin incision. A slit is made through the tunica, and the testis is freed.

4 The spermatic cord, which is attached to the testis, is doubly ligated, that is, two ties, close together, are made, and the cord is cut between the ties.

5 The other testis is similarly removed.

6 No thread sutures are necessary, the skin incision being merely closed with wound clips.

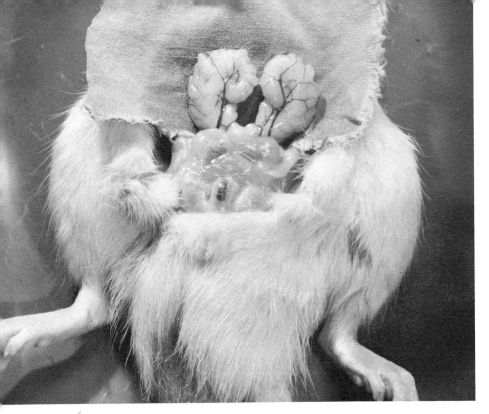

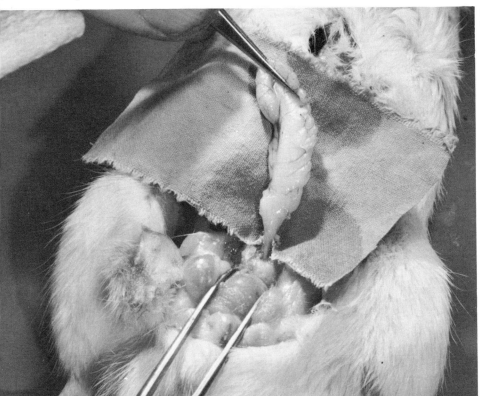

27 continued

Dissection of the seminal vesicles

1 The animals are killed with ether, and the lower abdominal region is freely exposed. Lying on either side of the bladder a frond-shaped structure will be seen. Closer inspection reveals that this consists of two parts—the seminal vesicle proper and the coagulatory gland. Both narrow down to thin stalks at their attachment to the vas deferens.

2 The two parts need not be separated from each other; both are dissected down to the stalk, and gentle pulling breaks the stalk. They are laid between filter papers moistened with saline. If a number are dissected at the same time, they are weighed together and the average weight per pair is calculated.

RESULTS OF TYPICAL EXPERIMENT

	Mg.
Control group	375
Castrate, noninjected group	80
Castrate, injected group	525

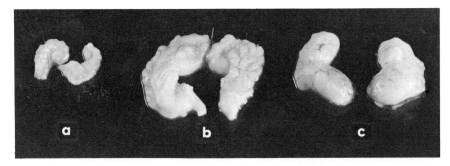

Seminal vesicles from (*a*) noninjected castrate, (*b*) injected castrate, (*c*) normal.

STUDENT'S REPORT

EXPERIMENT 28

MODE OF ACTION OF GONADOTROPIC HORMONES

In this experiment the fact is demonstrated that the action of gonadotropic hormones in stimulating accessory sex organs is indirect, i.e., by way of the gonads. Chorionic hormone, either from pregnancy urine or from pregnant mare serum, is used, rather than anterior lobe gonadotropin, as the latter is relatively ineffective in stimulating seminal vesicle growth.

MATERIALS AND EQUIPMENT

Sterile pack and surgical instruments.

Gonadotropic hormone preparation.

Antuitrin S (Parke-Davis) 500 I.U./cc or Follutein (Squibb) 1,000 I.U./cc.

PROCEDURE

Four groups of rats are studied, the operative and injecting work being divided among the students. The groups are as follows:

Group 1. Immature females (30 days old) are injected daily for 3 days with a total dose of 20 I.U. of gonadotropic hormone and killed at the end of 100 hours following the first injection. The vagina will be found to have opened, and a positive estrus smear, as described in Part I, Experiment 26, can be obtained. Enlargement and distention of the uterus will also be noted.

Group 2. This group is ovariectomized, as previously described, at the age of 21 days, and a week or 10 days later is given the same course of injections as Group 1. Examination at the end of 100 hours will show none of the maturation effects seen in Group 1, proving that the gonadotropic hormone produced its effect by way of the ovaries.

Group 3. This group consists of young males, 30 days old. They are injected in the same way as the previous groups and killed at the end of 100 hours. The seminal vesicles are dissected out and weighed.

Group 4. These animals are castrated as described in the previous experiment, when about 21 days old, given a week or so for recovery, and then injected as above. Seminal vesicles are dissected out and weighed. The increased weight of the seminal vesicles of Group 3 as compared to Group 4 is thus shown to be due to the stimulating action of gonadotropic hormone upon the testes.

RESULTS OF TYPICAL EXPERIMENT

	Vagina	Smears	Av. weight of uterus (mg.)
Group 1	Open	Estrus	100
Group 2	Closed	.....	11
	Av. weight of seminal vesicles (mg.)		
Group 3	22		
Group 4	5		

STUDENT'S REPORT

EXPERIMENT 29

MUSCLE-NERVE PHYSIOLOGY

While the animal of choice for work on muscle-nerve physiology is, of course, the frog, rats may well be used instead when they are more conveniently available. In this and the following experiment technics for such studies are described; many additional observations may be made on the same preparation.

MATERIALS AND EQUIPMENT

Inductorium.

Signal magnet.

Surgical instruments.

Setup for recording muscle contractions.

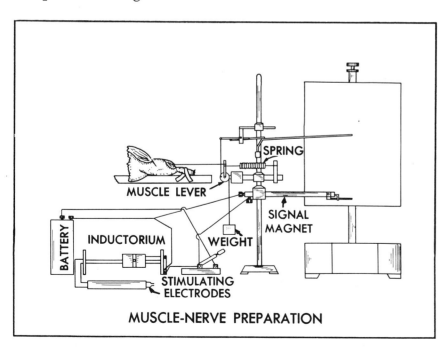

MUSCLE-NERVE PREPARATION

PREPARATION

The preparation used consists of the sciatic nerve and gastrocnemius muscle in situ. (The nerve supplying the gastrocnemius is the tibial, which, with the peroneal, forms the sciatic.)

1 With the animal lying on its side, make a skin incision parallel to the femur, along the lateral surface of the thigh and extending from a point about 2 cm. headward from the base of the tail about halfway to the knee.

2 Retract the skin. A groove will be seen posterior to the femur and running parallel to it. This groove lies between two muscles, the biceps femoris and the vastus lateralis.

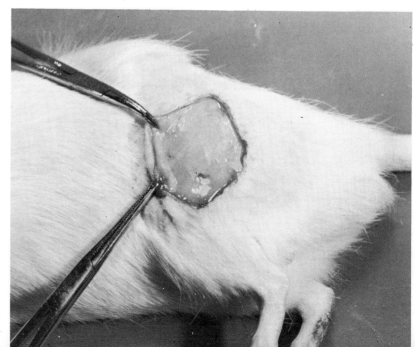

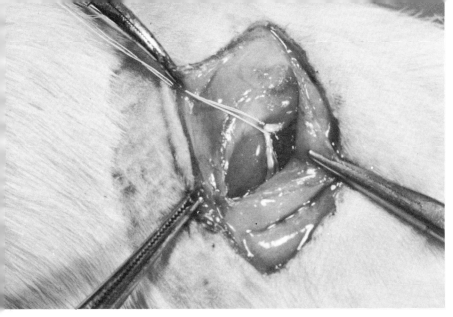

3 Deepening the groove by separation of the muscles exposes the sciatic nerve. It is a large nerve and easily freed for a considerable distance. Pass a loop under it and close the wound with clips.

4 Now free the tendon of Achilles from the ankle joint. Loosen the gastrocnemius from the skin and from surrounding muscles for some distance forward so that it can contract freely. Tie a thread tightly around the tendon.

5 Move the animal to the proper position in reference to the recorder. Tie the thread from the tendon to the recorder.

6 Return to the sciatic. Tie the thread tightly and cut centrally to the tie. Holding the nerve by the thread, you are now in position to stimulate the nerve. Keep the preparation warm and moist with Ringer-Locke solution.

PROCEDURE

Properties of muscle

1 Muscle twitch. With the apparatus properly adjusted and the drum running, close the circuit. This is the "make" shock. Then open the switch ("break" shock). Note that the "break" response is stronger than the "make."

29 continued

2 Treppe. Stimulate the nerve rapidly so that each stimulus comes immediately at the end of relaxation from the previous one. The strength of muscular contraction increases progressively, giving a staircase effect.

3 Tetanus. Loosen the interruptor screw of the inductorium as much as possible so that the frequency of interruptions will be slow. Stimulate. The impulses flow to the muscle too frequently to allow complete relaxation; the muscle is in partial tetanus, as shown by the sawtooth effect. Now tighten the interruptor screw so that the rate of interruptions is rapid, and stimulate. No time is permitted for relaxation; the muscle is in complete tetanus.

4 Influence of load. Fasten the ankle joint to the board on which the animal is lying by means of a thumb tack and the thigh by means of adhesive tape. Attach the weight pan to the thread running over the pulley. Use a shock which gives a response only at "break" and record the contractions as weights are added. Determine the maximum load capable of being lifted and determine the optimum load, i.e., the load at which the work done (weight × distance) is a maximum.

Action of drugs

1 Blocking agents. Slit the sheath of the nerve and let a drop or two of 50 per cent alcohol fall on the nerve fibers. After a few minutes the nerve will no longer conduct, even with strong stimulation.

2 Action of curare. As the blocking effect of alcohol is prolonged, another preparation must be used for this experiment. Have the femoral vein cannulated for curare injection.

3 With the nerve sheath slit, place a drop or two of curare solution (1 unit per cc.) on the nerve fibers. In contrast to alcohol, the nerve will continue to conduct and the muscle to contract.

4 Now inject 1 unit per kilogram into the jugular. Within a few seconds, stimulation of the nerve will no longer produce a response.

5 Now stimulate the muscle itself; it responds to stimulation. It is therefore evident that the action of curare was not on the

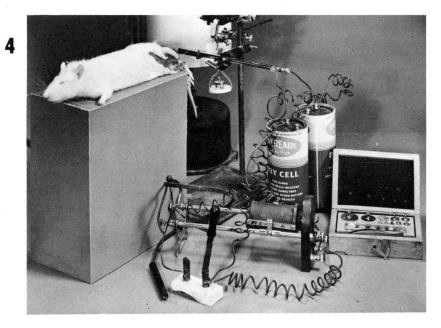

4

nerve fibers or on the muscle itself, but somewhere between the two, namely, at the myoneural junction.

Velocity of nerve impulse

The time elapsing between stimulation of a nerve and response of the muscle is determined with the stimulus applied, first, as far distally along the nerve as possible and, second, as far proximally as possible. This difference is the time required for the nerve impulse to traverse the length of the nerve between the stimulated points, and from it the velocity may be calculated. Since this time difference is very short, differences in muscle tension, friction in the apparatus, strength of stimulation, and extraneous movements of any sort will affect the outcome. The results are therefore not very reliable, but the experiment is included as a demonstration of the method used.

Special equipment

Two special items of equipment are necessary: a tuning fork, to which a short bristle is attached, and a special kymograph drum holder ("long paper drum"), in which the axle is mounted on ball bearings and which can be rotated rapidly. The kymograph drum is sooted, since it is not possible to use ink-writing instruments with the tuning fork.

PROCEDURE

1 Arrange the setup as follows: The thread which will be tied to the tendon of the gastrocnemius is passed over a pulley and downward to a muscle lever. The writing point of the lever is placed directly over the writing point of a signal magnet. The latter is connected in series with the inductorium, which is set to give single shocks; when the switch is closed, the inductorium and signal magnet react simultaneously.

2 Free the sciatic nerve and the attachment of the gastrocnemius muscle as previously described. Trace the branches of the sciatic into the popliteal fossa; the large medial branch is the tibial, which supplies the gastrocnemius.

3 Open the sheath of the sciatic and separate the tibial nerve as far bodyward as possible, tie a thread around it, and cut bodyward to the thread. Keep the nerve in place, covered by muscle, as much as possible.

4 Now move the animal into position for recording muscle contractions, tying the thread tightly to the tendon. With adhesive tape fasten the foot to the edge of the board on which the animal lies and adjust the position of the board so that tension is exerted on the muscle and the lever is held suspended by the thread.

5 One student will now expose the tibial nerve and lay it across the electrodes, as far distally as possible. With the writing points touching lightly against the drum and in a vertical line, another student now spins the drum. A third student sets the tuning fork to vibrating and, as a fourth student closes the switch, places the hair on the tuning fork against the drum just below the line being written by the signal magnet.

29 continued

RESULTS—SAMPLE RECORDS

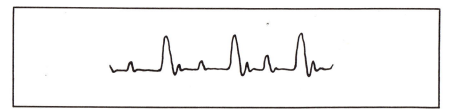

Simple twitch, showing stronger contraction at "break."

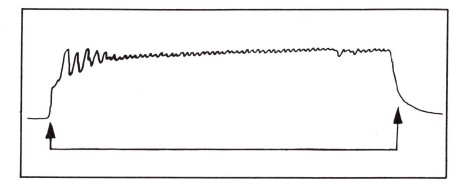

Partial tetanus.

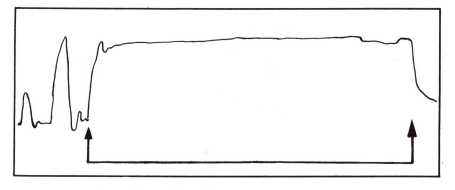

Complete tetanus.

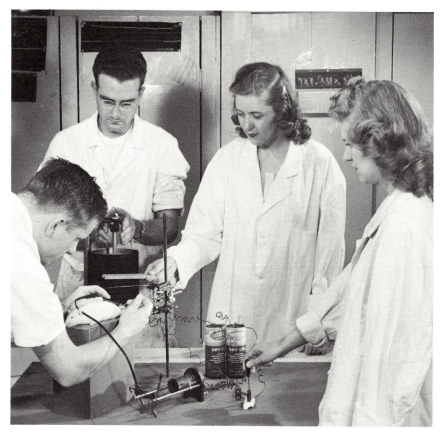

6 The drum is stopped, its position changed, and the experiment repeated, but with the electrodes placed at the extreme proximal end of the nerve. Finally, the nerve may be wrapped in cotton soaked in ice-cold Ringer solution and the stimulus again applied to the proximal end, to show the effect of cold on the velocity.

7 The length of the nerve between the points stimulated is now measured and the calculations are made.

CALCULATIONS

In the graph shown, the tuning fork vibrated 320 times per second; a wave is therefore equivalent to 3.1 sigma. The difference in time interval between the upper left hand (distal stimulation) and the upper right hand (proximal stimulation) is about one-sixth wave or about 0.5 sigma. The length of the nerve was 3 cm.; therefore, the velocity was about 60 m/sec. Cooling the nerve reduced the velocity to about 2.5 m/sec.

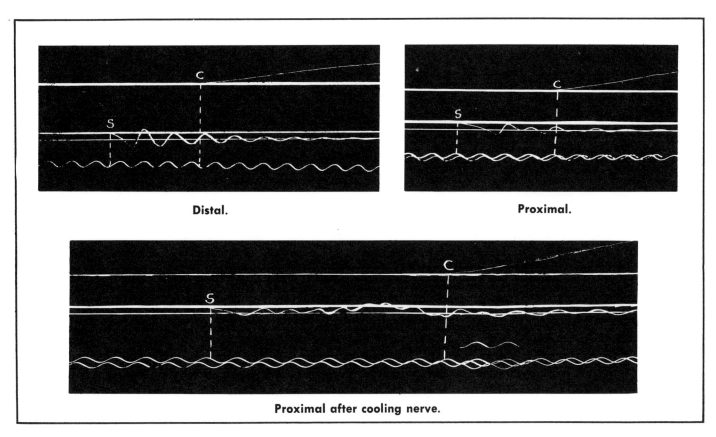

Record of "Nerve Impulse Velocity" experiment. S, signal; C, contraction.

 continued

STUDENT'S REPORT

EXPERIMENT 30

THE SOURCE OF ENERGY FOR MUSCLE CONTRACTION

In this experiment the fact that muscle glycogen is the source of energy for contraction is demonstrated by comparing the glycogen content of a gastrocnemius muscle, stimulated to the point of fatigue, with that of the opposite, resting, muscle.

MATERIALS AND EQUIPMENT

Setup for muscle contraction.
Reagents for glycogen determination:
    30 per cent potassium hydroxide.
    95 per cent alcohol.
Reagents for Shaffer-Somogyi method for sugar determination (see Part II, Exper. 14).

PROCEDURE

1 Prepare the rat as in the previous experiment.

2 With the muscle pulling against a fairly strong spring tension or a 50-gm. weight, stimulate at a rate of twice per second, until the response becomes definitely weaker. This will require 30–45 minutes.

3 Then quickly dissect out the gastrocnemius muscles on both sides and proceed with the glycogen determination.

4 Weigh the muscles as rapidly as possible and place in 15-cc. test tubes. Add approximately 2 cc. of 30 per cent potassium hydroxide and heat in a boiling water bath until the tissue is in solution (30 min. to 1 hour).

5 Precipitate the glycogen with 1.5–2 volumes of 95 per cent alcohol, heat the tubes to boiling, and cool to room temperature.

6 Centrifuge. (The centrifuging may be postponed until the following day if necessary.)

7 Discard the supernatant fluid, add 1 cc. of 1 N sulfuric acid and hydrolyze by heating on a water bath for 2–3 hours.

8 Dilute the sample from the resting muscle to 25 cc. and use a 5-cc. aliquot for analysis. The entire sample from the fatigued muscle may be used.

9 Analyze for glucose by the Shaffer-Somogyi method as described in Experiment 34.

CALCULATIONS

The milligrams of glucose multiplied by 0.95 give the milligrams of glycogen.

Average results obtained by this method run from 600 to 750 mg/100 gm of resting muscle and from 150 to 250 mg. for fatigued muscle.

30 continued

STUDENT'S REPORT

EXPERIMENT 31

THE BELL-MAGENDIE LAW OF SPINAL NERVE ROOTS

The purpose of this experiment is to demonstrate the fact that the dorsal nerve roots are afferent only and that the ventral nerve roots are efferent only (the Bell-Magendie law).

MATERIALS AND EQUIPMENT

Inductorium.

Usual surgical instruments.

Bone-cutters—a pair of fingernail clippers is satisfactory.

PREPARATION

In this experiment the rat must be kept deeply anesthetized during the operation but the anesthetic should be lightened somewhat while the nerve roots are being stimulated. Use Ringer-Locke rather than saline solution to keep nerve tissue warm and moist.

PROCEDURE

1 Place the animal in the prone position, with the pelvis lying over a small bottle filled with warm water. This serves to maintain body temperature and prevents interference with respiration and circulation because of pressure against the animal's body.

2 Make a mid-line incision from the second to the sixth lumbar vertebrae and extend the incision laterally at each end toward the right. Retract the skin; this exposes the silvery lumbodorsal fascia covering the vertebral column. Make an incision just

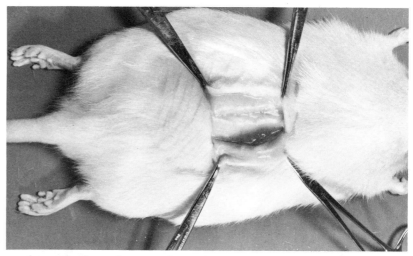

to the right but close to the spinous processes, another just to their left, and another lateral to the right incision and parallel to it, just lateral to the ends of the articulate processes.

3 By rapid dissection, clear out the muscle lying between the two right incisions.

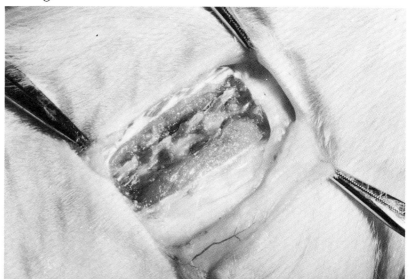

31 continued

4 With the bone-cutters cut off the articulate processes as fully and deeply as possible; then, starting with the sixth and extending to the second, cut off the spinous processes as well. Great care must be exercised not to injure the spinal cord or nerves while this is being done. Take very small bites with the cutters through the lamina on each side and then lift up the spinous process. The view of the exposed cord is improved by nibbling away the stumps of the overhanging articulate processes. The spinal cord is inclosed within the theca, which is now opened by picking up with fine-pointed forceps and slitting with fine scissors.

5 Expose the fifth dorsal root ganglion. The ganglion lies a little headward of the articulate process and close to the theca.

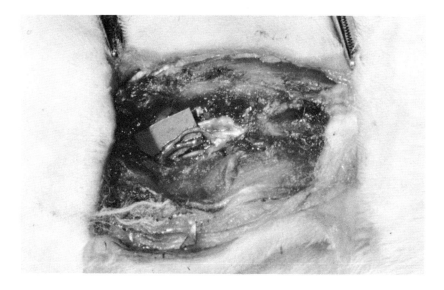

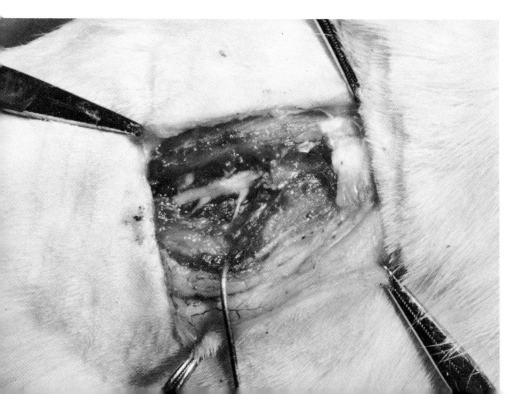

6 By placing a thread under the dorsal root and lifting gently, the branching of the spinal nerve into the dorsal and ventral roots can be seen.

7 Expose the fourth nerve in the same manner, and the preparation is now ready for stimulation.

8 Bring the thread around the dorsal root of the fifth as far outward toward the ganglion as possible, tie, and cut distal to the tie. Stimulate the central end. Movements of the abdominal muscles, psoas, and leg will be seen. It is therefore afferent.

9 Place a thread around the dorsal root of the fourth, but tie as close to the cord as possible and cut between the cord and the tie. Stimulate the peripheral end. No movement will be seen. The dorsal root is therefore *only* afferent.

10 Repeat the above procedures on the ventral roots, stimulating the central end of the fifth and the peripheral end of the fourth.

No response will follow central stimulation, while violent contractions of the leg muscles occur when the peripheral end is stimulated. The ventral root is therefore only efferent.

STUDENT'S REPORT

EXPERIMENT 32

STIMULATION OF THE MOTOR AREAS OF THE CEREBRAL CORTEX

MATERIALS AND EQUIPMENT

Usual surgical instruments.

Small drill.

Bone-cutters.

Inductorium.

PREPARATION

Anesthetize the rat with 40 mg. Nembutal per kilogram, reinforce with ether during the operation, but keep anesthesia light while cortex is being stimulated. Use Ringer-Locke rather than saline.

PROCEDURE

1. Place the rat in the prone position, make a mid-line incision through the skin of the head and retract to either side. Clear away connective tissue over the skull.

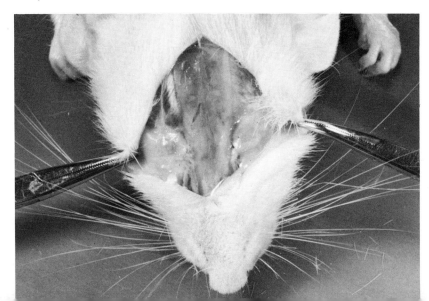

2. Drill three small holes, quite close together, through the anterior region of the right frontal bone. Avoid the mid-line, as the superior sagittal sinus lies directly below. As the drill-hole deepens, lighten the pressure to prevent forcing the bit into the brain.

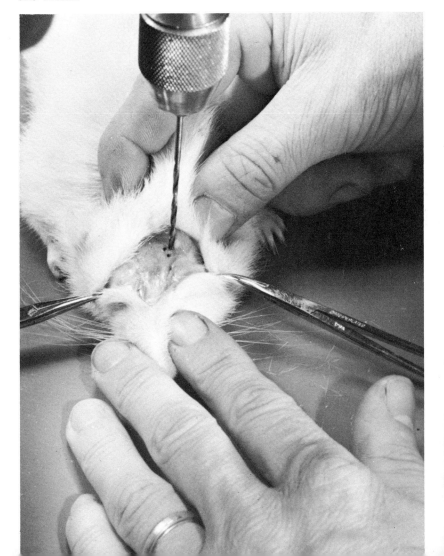

3 Clip the bone carefully away between the holes and enlarge the orifice thus made at its margins until the frontal lobe, between the inferior and superior cerebral veins, is exposed. There will be some hemorrhage from the skull; this can be controlled by rubbing in small amounts of warm bone wax. (One-third beeswax and two-thirds paraffin.)

4 Carefully clip away the dura, exposing the surface of the brain.

5 With the points of the stimulating electrodes set close together, explore the exposed surface of the brain with light, intermittent shocks. The opposite side of the animal should be carefully watched for responses. Several areas of the motor cortex can be fairly sharply delineated: an area serving the nose, ears, and whiskers; another for the neck and forelimbs; and another for the hind limbs and tail.

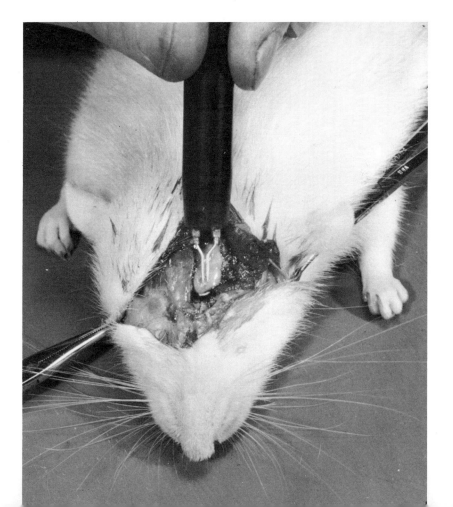

32 continued

STUDENT'S REPORT

EXPERIMENT 33

EFFECTS OF AUTONOMIC DRUGS ON THE CIRCULATION

In this and the following experiment the action of certain drugs acting upon the autonomic nervous system is studied. As an example of an agent which "blocks" postganglionic cholinergic nerve effects, the action of atropine in abolishing the results of vagal stimulation is shown. The action of acetylcholine itself and the ability of physostigmine (eserine) to potentiate its action is demonstrated, and pilocarpine is used to duplicate the muscarinic action of acetylcholine. Adrenalin is used as an example of the sympathomimetic drugs.

MATERIALS AND EQUIPMENT

Blood pressure setup.

Inductorium.

Solutions of: atropine sulfate . . . . 2 mg/cc
acetylcholine bromide . . . . 0.02 mg/cc
physostigmine salicylate . . . . 0.02 mg/cc
pilocarpine nitrate . . . . 0.5 mg/cc
adrenalin hydrochloride . . . . 0.003 mg/cc

PROCEDURE

Two rats are used in this experiment, the first for the work on atropine and the vagus, the other for the remaining drugs.

1 Cannulate the left carotid and the right femoral.

2 Free the vagus on the right side.

3 Connect the carotid to the tambour and, while a normal blood pressure record is being taken, stimulate the peripheral end of the vagus with weak shocks from the inductorium. The strength of stimulus should be such that only a slowing of the heart and a significant, but not large, drop in the blood pressure result. Leave the inductorium at that setting.

4 When the record has returned to normal, inject 2 mg/kg of the atropine sulfate solution via the femoral vein. Quickly repeat the vagal stimulation. No effect will now be produced, because of the "blocking" action of the atropine.

5 Cannulate a carotid and the opposite femoral of another rat. Attach the arterial cannula to the tambour and take a normal record.

6 Inject via the femoral 0.02 mg/kg of the acetylcholine solution.

7 Let the blood pressure return to normal and then inject 0.02 mg/kg of the physostigmine solution.

8 Again let the blood pressure return to normal. Mix equal portions of the two solutions and inject 1 cc/kg of the mixture. (This contains only half as much of each drug as previously used.) It will be found that the effect, in spite of the reduced

33. continued

dosage, is greater and more prolonged, owing to the potentiating action of physostigmine.

9 When the blood pressure has returned to normal, inject 0.5 mg/kg of pilocarpine. In addition to the blood pressure effect, observe other effects, such as salivation, lacrimation (the tears are blood-tinged), and respiratory difficulty due to bronchial spasm.

10 The action of adrenalin has previously been studied but may be repeated here, using a dose of 0.005 mg/kg, in order to complete the study.

The experiment may be extended by demonstrating the blocking action of atropine against the effects of acetylcholine and pilocarpine and also the mydriatic effect of atropine upon the pupil of the eye.

RESULTS—SAMPLE RECORDS

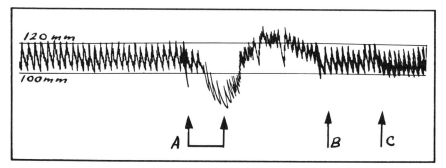

The action of atropine in abolishing the effect of vagal stimulation. At *A* the vagus was stimulated; at *B* atropine was injected, and the vagal stimulation was repeated at *C*.

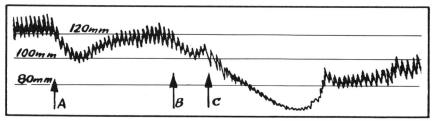

Acetylcholine (0.02 mg/kg) was injected at *A*, followed by 0.02 mg/kg of physostigmine at *B*. A mixture of the two solutions was injected at *C*.

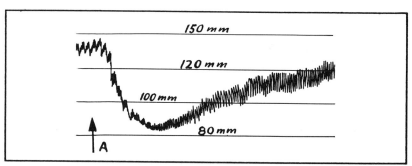

Record showing the effect of the injection of 0.5 mg/kg of pilocarpine.

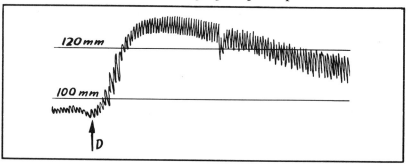

Record showing the effect of 0.005 mg/kg of adrenalin on the blood pressure and on the heart.

STUDENT'S REPORT

EXPERIMENT 34

EFFECTS OF AUTONOMIC DRUGS ON GASTRIC AND INTESTINAL MOTILITY

MATERIALS AND EQUIPMENT

Tambour and recording apparatus as described in Part I, Experiment 17.

Solutions of: adrenalin, 0.005 mg/cc.
acetylcholine, 0.05 and 0.5 mg/cc.
physostigmine, 0.05 mg/cc.
pilocarpine, 0.1 and 1 mg/cc.

PROCEDURE

Two rats will be required, one for studies on the stomach, the other for the intestine. Both should have a jugular cannulated, for ease of injection of the solutions.

1 Prepare a rat for recording gastric motility as described in Part I, Experiment 17.

2 Take a normal tracing, then inject, via the femoral, 0.005 mg/kg of adrenalin. An immediate relaxation of the stomach follows, which is of short duration.

3 When recovery has occurred, inject an acetylcholine solution. The effect of acetylcholine, similar to vagal stimulation, is not always predictable. If the dose used is small, 0.05 mg/kg, there is usually a fall; if it is higher, 0.5 mg/kg, a rise in tonus usually results.

4 Wait until recovery, then inject a small dose of physostigmine, 0.05 mg/kg. This will have little effect.

5 Then combine the physostigmine with the dose of acetylcholine previously given. The effect, in either direction, will be accentuated and prolonged.

6 Finally, after the normal condition has again been reached, inject pilocarpine in a dose first of 0.1 mg/kg and then of 1 mg/kg. Pilocarpine always produces a distinct increase in motility.

7 Repeat the above experiment on the second rat, the intestine having been cannulated as described in Part I, Experiment 17. The effects on the intestine will be found, in general, to duplicate those on the stomach.

RESULTS—SAMPLE RECORDS

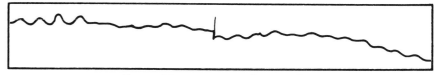

Normal tonus waves of stomach.

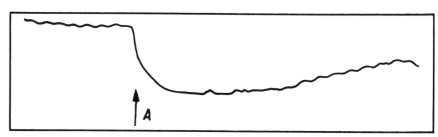

Relaxation produced by 0.005 mg/kg of adrenalin.

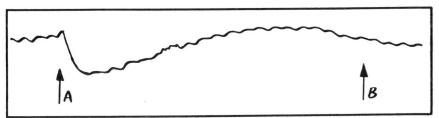
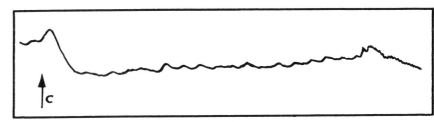

Acetylcholine (0.05 mg/kg) was injected at *A*; 0.05 mg/kg of physostigmine at *B*; a mixture of the two solutions was given at *C*. Note the longer duration of the depressant effect.

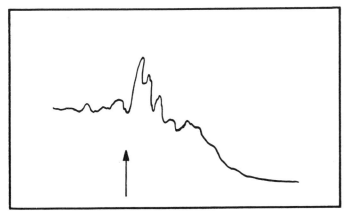

Effect of a larger dose (0.5 mg/kg) of acetylcholine on gastric motility.

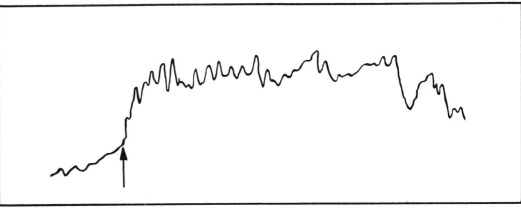

Effect of 0.1 mg/kg of pilocarpine on gastric motility.

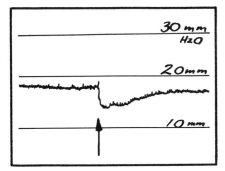

Effect of 0.005 mg/kg of adrenalin on intestinal motility.

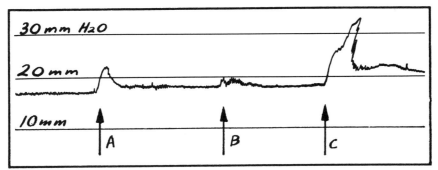

Acetylcholine (0.05 mg/kg) was given at *A*, followed by 0.05 mg/kg of physostigmine at *B*. A mixture of both solutions was injected at *C*.

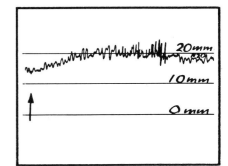

Effect of 1 mg/kg of pilocarpine on intestinal motility.

34 continued

STUDENT'S REPORT

PART II SPECIAL EXPERIMENTS

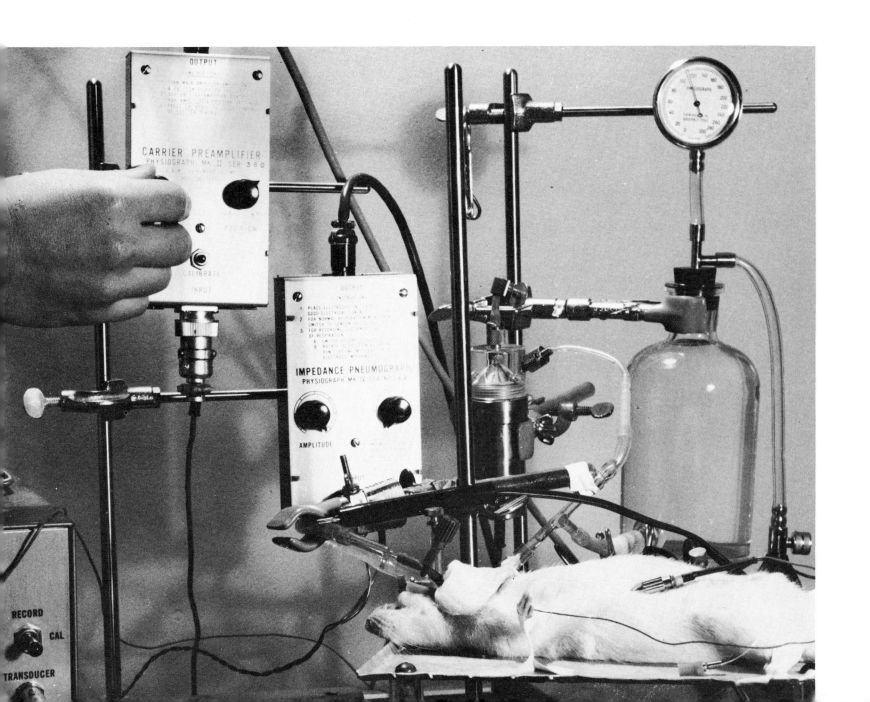

LIST OF EXPERIMENTS Part II

This group of more advanced experiments is intended for use in a second semester of physiology if the course lasts for a full year. It also gives the instructor a choice of experiments to assign the brighter students in a class or the research-minded individual. All or most of the preceding group should have been performed before venturing into this section. This group includes the following experiments:

BLOOD

1. Hemoglobin Determination, Van Slyke Method
2. Reaction of Carbon Monoxide with Hemoglobin: Difference between Carbon Monoxide and Cyanide Poisoning
3. Determination of the Freezing-Point of Blood Plasma

HEART AND CIRCULATION

4. Effects of Carbon Dioxide, Lactic Acid, and Asphyxia on the Blood Pressure
5. Effect of Rupture of the Aortic Valves on the Blood Pressure (Water-Hammer Pulse)
6. Lateral versus Thrust Blood Pressure
7. Blood Velocity
8. Circulation Time
9. Determination of the Cardiac Output
10. Production of Renal Hypertension

RESPIRATION

11. Artificial Respiration by the Drinker Respiratory Chamber Method
12. Determination of Oxygen Consumption, Respiratory Quotient, and Basal Metabolic Rate

DIGESTION AND METABOLISM

13. The Bromsulphalein Test for Liver Function
14. The Rate of Glucose Absorption. (Cori and Cori Coefficient)
15. Effect of Adrenalin on Blood Sugar
16. Effect of Insulin on Blood Sugar
17. Alloxan Diabetes

EXCRETION

18. The Phenolsulfonephthalein Test for Kidney Function

ENDOCRINE SYSTEM

19. The Aschheim-Zondek Pregnancy Test
20. The Effects of Long-continued Estrin Administration
21. Effects of Hypophysectomy on Weight Gain and the Estrus Cycle

NERVOUS SYSTEM

22. Influence of the Sympathetics on the Blood Vessels of the Ear
23. A Study of the Spinal Reflexes
24. The Production of Decerebrate Rigidity
25. Effects of Labyrinthectomy
26. Establishment and Extinction of a Conditioned Reflex

ANTIBODY FORMATION

27. Anaphylactic Shock

ISOTOPES

Radiation Protection Procedures and Laboratory Regulations

28. Utilization of C^{14} Labeled Glucose in the Rat
29. Intestinal Absorption and Tissue Uptake of Co^{60}-Vitamin B_{12}

ENZYMES

30. Transaminase Levels and Tissue Damage
31. Biosynthesis of Urea

EXPERIMENT 1

HEMOGLOBIN DETERMINATION, VAN SLYKE METHOD

In this method the hemoglobin content of the blood is determined by saturating the blood with oxygen, releasing the oxygen in the apparatus, and measuring its volume. The hemoglobin is computed on the basis of 1 gm. of hemoglobin combining with 1.36 cc. of oxygen.

MATERIALS AND EQUIPMENT

Solution A.

Neutral saponin-ferricyanide. 6 gm. $K_3Fe(CN)_6$ and 3 gm. of saponin are dissolved in 1 liter of distilled water.

Solution B.

0.1 N lactic acid (1 cc. conc. lactic acid plus 99 cc. H_2O). Equal volumes of solutions A and B are mixed just before using. Their function is to lake the blood.

1 N sodium hydroxide.

Used to absorb carbon dioxide (4 gm. NaOH/100 cc.).

2 gm. sodium hydrosulfite ($Na_2S_2O_4$) and 0.2 gm. sodium anthraquinone-beta-sulfonate are weighed out and mixed dry.

Just before using, dissolve in 10 cc. N potassium hydroxide and filter rapidly through cotton. This solution is used to absorb oxygen (5.6 gm KOH/100 cc.).

Caprylic alcohol.

1 250-cc. Florence flask.

Water bath at 37° C.

2 10-cc. syringes with short rubber tubes and clamps attached.

1 1-cc. syringe with short rubber collar over tip.

Instruments for carotid cannulation.

Mineral oil.

Van Slyke gasometric apparatus.

The manometric apparatus is used in the experiment as here described; the volumetric apparatus may, however, be used.

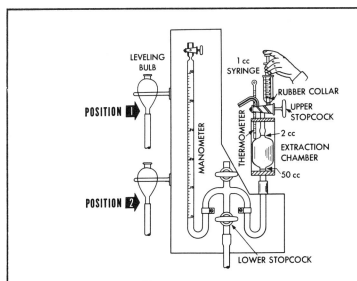

INSTRUCTIONS FOR USE

The figure shows the leveling bulb in three positions, marked *1, 2,* and *3*. These are the positions referred to in the further discussion.

If the bulb is in position *1*, the mercury will flow into the cup from the chamber when upper and lower stopcocks are open. In position *2* the flow will be from cup into chamber. Position *3* is used with the lower stopcock open and the upper closed to bring the solution down into the extraction chamber. Never open the upper stopcock with the bulb in position *3* with the lower stopcock open, or the solution and mercury will be drawn over into the manometer. The student should practice shifting the position of the bulb and noting its effect on the flow of mercury before undertaking the experiment.

 continued

PROCEDURE

1 Cannulate the carotid artery and inject 0.25 cc. of heparin. Mix as usual by repeated withdrawals and reinjections. Withdraw about 5 cc. of blood.

2 Place about 3 cc. of blood in a 250-cc. Florence flask, cork, and rotate in a water bath at 37° C. for at least 15 minutes.

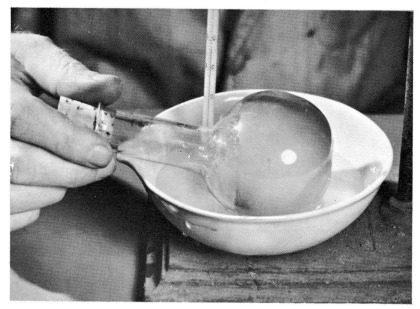

This is for the purpose of saturating the blood with oxygen; therefore rotate the flask so that the blood flows in a thin layer, with as much surface exposed as possible.

3 While this is being done, the solutions to be used must be freed of air. Starting with the cup full of mercury, place the bulb in position 2 and let the mercury flow down almost to the bottom of the cup. Pour about 10 cc. of 1 N NaOH into the cup and let nearly all of it flow into the chamber. Place a few drops of mercury in the cup and let a little of it fill the bore of the stopcock. This is to make a mercury seal. Remove the excess NaOH from the cup. With the bulb in position 3, lower the mercury until its meniscus stands at the 50-cc. mark, close the lower stopcock, and shake for 3 minutes. Notice the bubbles of air escaping from the solution.

4 The solution is now air-free and must be collected under oil. Draw about 2 cc. of oil into a 10-cc. syringe which has attached to it a short piece of rubber tubing. Force a layer of oil into the cup over the mercury seal, leaving some oil in the syringe. With the bulb in position 1, let the solution rise slowly into the cup. Draw off most of the solution into the syringe, clamp the tube, and place in a test-tube clamp on a ring stand until ready to use.

5 Make up the sodium hydrosulfite solution as described under "Materials and Equipment." Place the solution in the chamber, seal, and deaerate as described above. Withdraw under oil.

6 The last solution to be deaerated is the laking solution. Measure 5 cc. each of solutions A and B and mix. Before the mixture is introduced, the chamber should be rinsed with dilute lactic acid, as the previous solutions were strongly alkaline. Pour 7.5 cc. of the A–B mixture into the cup and place about 4 drops of caprylic alcohol on top. Lower into the chamber, seal, and deaerate for 3 minutes. Now raise the solution and let 6 cc. flow into the cup, leaving 1.5 cc. in the chamber.

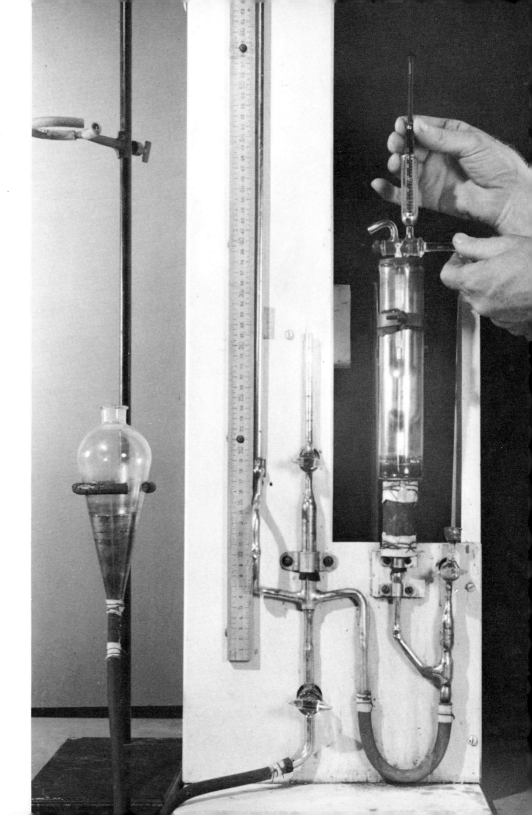

7 Measure exactly 1 cc. of blood from the Florence flask into a syringe with a rubber collar on the tip. Place the tip directly over the bore of the stopcock at the bottom of the cup. With bulb in position 2 and lower stockcock open, now open upper stopcock, and suction will pull the blood sample into the chamber. Wash through with 1 cc. of the laking solution which has remained in the cup. Make a mercury seal and remove excess solution from the cup. Lower the solution to the 50-cc. mark and shake for 3 minutes. Raise the solution to the 2-cc. mark and read the manometer. This reading is called "P_0" and represents the pressure of all the gases liberated from the blood. Also read the temperature at this point.

8 Lower the solution about 5 cc. and pour 2 cc. of the deaerated NaOH into the cup. Do this under oil to prevent reabsorption of air. Then draw into the chamber, very slowly, drop by drop, exactly 1 cc. of the sodium hydroxide solution, taking at least 30 seconds for this operation. Now bring the meniscus of the solution exactly to the 2-cc. mark and read the manometer. This reading is P_1. It represents the pressure of the oxygen and nitrogen in the blood, the carbon dioxide having been absorbed by the NaOH. Therefore, $P_0 - P_1 = CO_2$ pressure.

9 Having removed the excess NaOH from the cup, pour into it 2 cc. of the hydrosulfite, under oil. Permit 0.5 cc. to enter the chamber very slowly. Let the solution rise to the upper stopcock to compress the air and then let an additional 0.5 cc. of the hydrosulfite enter. Take 2 minutes for the total absorption time. Now bring the solution exactly to the 2-cc. mark and

 continued

read the manometer. This reading is P_2, and the difference between it and P_1 is the oxygen pressure.

10 Drain the solutions out of the apparatus. Wash with strong alkaline hydrosulfite solution. Rinse and fill the chamber with water until used again.

CALCULATIONS

Look up the appropriate factor in the table given at end of the experiment, which is taken from Peters and Van Slyke, *Quantitative Clinical Chemistry*, 2:325. Multiply this factor by the oxygen pressure in millimeters and divide by 1.36. This gives the result in grams of hemoglobin per 100 cc. of blood.

Temperature	24° C.
P_0	428 mm.
P_1	388 mm.
P_2	299 mm.
Factor	0.2414

$$P_1 - P_2 = 89 \text{ mm.}$$
$$89 \times 0.2414 = 21.48$$
$$\frac{21.48}{1.36} = 15.8 \text{ gm. Hb. per 100 cc. of blood.}$$

The factor 0.2414 is arrived at as follows: The amount of gas in a space of 2 cc., exerting a pressure of 1 mm. is $\frac{2 \text{ cc.}}{760}$ at 0° C.; at 24° it would be $\frac{2}{760} \times \frac{273}{297} = 0.002419$ cc. Since 1 cc. of blood was used, but the hemoglobin is to be expressed per 100 cc., the factor becomes 0.2419. The slight difference is a correction made for the weight of the solution on the mercury.

IRON METHOD

The iron content of the blood is determined by the colorimetric method of Wong, and the hemoglobin is calculated.

MATERIALS AND EQUIPMENT

Iron-free conc. sulfuric acid.

> Test the c.p. acid for iron by diluting 1 cc. of the acid to 20 cc. with distilled water and add 4 cc. of 3 N potassium sulfocyanate reagent; a red color indicates the presence of iron.

Sat. potassium persulfate solution.

10 per cent sodium tungstate solution.

Standard iron solution (0.1 mg. iron per cc).

> Dissolve 86 mg. of ferric ammonium sulfate in 10 cc. of distilled water; add 2 cc. of 10 per cent iron-free sulfuric acid and dilute to 100 cc.

3 N potassium sulfocyanate solution (29 gm. per 100 cc.).

50-cc. volumetric flask.

2 test tubes calibrated at 20 cc.

Colorimeter.

Instruments for carotid cannulation.

PROCEDURE

1 Obtain about 2 cc. of blood by carotid cannulation. Transfer 0.5 cc. to a 50-cc. volumetric flask and add 2 cc. of the iron-free conc. sulfuric acid. Agitate the mixture for 2 minutes. Add 2 cc. of saturated potassium persulfate and shake. Dilute to about 25 cc. with distilled water and add 2 cc. of 10 per cent sodium tungstate solution. Mix, cool to room temperature, and dilute to the 50-cc. mark. Filter through dry paper into a dry beaker.

2 Transfer 20 cc. of the filtrate into a test tube calibrated at 20 cc. Into a similar test tube place 1 cc. of the standard iron solution. Add 0.8 cc. of iron-free conc. sulfuric acid and dilute to the 20-cc. mark with distilled water. Cool to room temperature. Add to each tube 1 cc. of saturated potassium persulfate and 4 cc. of 3 N potassium sulfocyanate solution. Mix and compare in a colorimeter.

CALCULATIONS

The colorimeter reading of the blood solution was 19.5, with the standard at 20. The 25 cc. of final standard solution contained 0.1 mg. of iron; therefore, the 25 cc. of the final blood solution contained:

$$\frac{0.1 \times 20}{19.5} = 0.103 \text{ mg.}$$

This 25-cc. final solution represented 0.2 cc. of blood; therefore, 100 cc. of blood contained:

$$0.103 \times \frac{100}{0.2} = 51.5 \text{ mg.}$$

Since it is established that the iron content of hemoglobin is 0.336 per cent, the amount of hemoglobin found in this experiment is:

$$\frac{51.5}{0.336} \times 100 = 15.3 \text{ gm. per 100 cc. of blood.}$$

FACTORS FOR CALCULATION OF O_2, CO, OR N_2 CONTENT OF BLOOD WHEN 1-CC. SAMPLE IS USED

Temperature (° C.)	Factor To Give Volume per Cent
15	0.2493
16	0.2485
17	0.2478
18	0.2468
19	0.2459
20	0.2450
21	0.2441
22	0.2432
23	0.2423
24	0.2414
25	0.2406
26	0.2398
27	0.2390
28	0.2382
29	0.2374
30	0.2366
31	0.2358
32	0.2350
33	0.2342
34	0.2333

 continued

STUDENT'S REPORT

EXPERIMENT 2

REACTION OF CARBON MONOXIDE WITH HEMOGLOBIN: DIFFERENCE BETWEEN CARBON MONOXIDE AND CYANIDE POISONING

This experiment demonstrates the property of carbon monoxide of combining with hemoglobin to prevent the normal transport of oxygen, as contrasted with cyanide, which does not interfere with oxygen transport.

MATERIALS AND EQUIPMENT

Instruments for cannulation.

Carbon monoxide generator (Fig. 1).

One-way valves.

Spirometer

0.05 per cent solution of KCN in saline (NaCN is equally satisfactory).

Reagents and apparatus for determining blood gases by Van Slyke method.

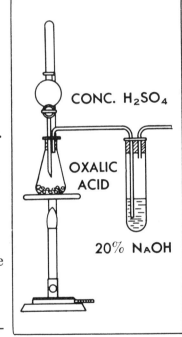

Fig. 1.—Carbon monoxide generating apparatus.

Preparation of carbon monoxide

To 5 gm. of oxalic acid add 10–15 cc. of conc. sulfuric acid by means of a separatory funnel. Heat the mixture until a vigorous reaction takes place. This should be done under a hood or with the laboratory windows open. Pass the gas generated through a 20 per cent solution of NaOH to remove the CO_2 and collect the CO in a large syringe. Prepare a mixture containing 1 per cent CO by transferring 20 cc. of CO to a spirometer and then adding 2,000 cc. of air.

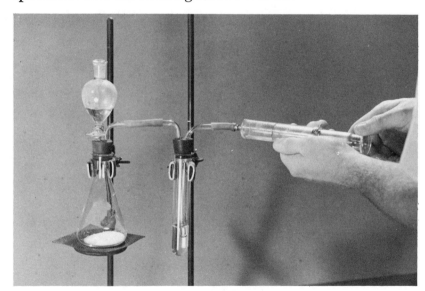

PROCEDURE, carbon monoxide experiment

In order to demonstrate the effect of the carbon monoxide on the hemoglobin, one sample of blood is analyzed previous to exposure to CO and one after exposure.

| Cannulate the trachea and carotid artery. Remove 1 cc. of blood, taking care to prevent exposure to the atmosphere, and

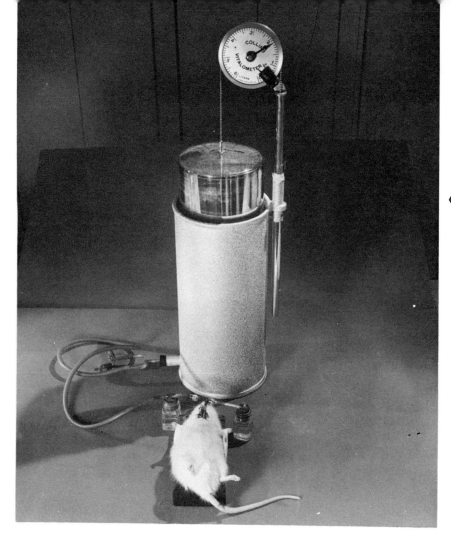

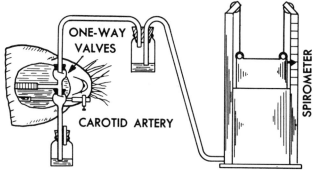

Hookup of rat to one-way valve system and to spirometer.

2 continued

analyze for oxygen by the procedure given in Part II, Experiment 1.

◁ **2** During the time that the normal sample of blood is being analyzed, the animal is exposed to carbon monoxide. Attach the tracheal cannula to the one-way valves as shown in the illustration and allow the animal to breathe the gas mixture for about 10 minutes. (Note the bright-red color which develops in the nose and feet.) Remove a second sample of blood and analyze for oxygen and carbon monoxide by the procedure given in Part II, Experiment 1, with the following additions:

3 After the oxygen has been absorbed and the level of the solution is being lowered to the 2-cc. mark, care must be taken not to overrun the mark. Carbon monoxide has a high affinity for reduced hemoglobin, and the manipulation required to bring the level back will cause increased absorption of the carbon monoxide. Record the P_2 reading and exhaust the remaining gases from the chamber. Bring the level to the 2-cc. mark; record the P_3 reading.

CALCULATIONS

Oxygen
$$\text{Vol. per cent } O_2 = (P_1 - P_2) \times \text{factor.}$$
Carbon monoxide
$$\text{Vol. per cent } CO = 1.024 \, (P_2 - P_3) \times \text{factor} - 1.2.$$

Factor for oxygen, nitrogen, and carbon monoxide is found in Peters and Van Slyke, *Quantitative Clinical Medicine*, **2**:325, or in table at the end of Part II, Experiment 1.

PROCEDURE, cyanide experiment

1 Cannulate the trachea, one of the carotid arteries, and the opposite jugular vein.

2 Introduce slowly, 0.05 cc. at a time, into the jugular vein a solution of KCN containing 0.5 mg. per cubic centimeter. When the animal shows signs of distress, as evidenced by rigidity and convulsions, remove 1 cc. of blood from the carotid and analyze for oxygen.

The lethal dose of KCN depends largely on the rate at which it is administered. If 0.05 cc. of the solution is injected at intervals of about 1 minute, the signs of distress described will usually be present when 1 cc. has been given.

It should be noted from the results that the oxygen content of the blood after exposure to carbon monoxide is much lower than the normal value but that after cyanide poisoning the blood is still carrying its normal amount of oxygen.

RESULTS OF TYPICAL EXPERIMENT

Vol. per cent of oxygen in first sample (before breathing CO) = 18.0.
Vol. per cent of oxygen after breathing air containing 1 per cent CO for 10 minutes = 6.0–10.0.
Vol. per cent of oxygen after KCN injection = 18.0.

STUDENT'S REPORT

EXPERIMENT 3

DETERMINATION OF THE FREEZING-POINT OF BLOOD PLASMA

The determination of the freezing-point is usually carried out in a Beckmann apparatus, which consists essentially of a rather expensive differential thermometer. In the procedure described in this experiment the ordinary apparatus found in a physiology laboratory has been substituted for the thermometer and found to give satisfactory results. There is an added advantage in that a permanent record is obtained which shows all the changes in temperature that take place during the freezing process.

MATERIALS AND EQUIPMENT

Kymograph.

Recording assembly, consisting of tambour and sooted drum.

Thermos flask containing an ice-alcohol freezing bath.

Metal capsule similar to those used on the heat indicator for automobiles.

Sample tubes—inside diameter slightly larger than the outside diameter of the metal capsule.

Instruments for carotid cannulation.

Centrifuge and tubes.

Ethyl chloride.

Distilled water (for extreme accuracy water should be triple-distilled).

White shellac (may be thinned with methyl alcohol).

PROCEDURE

1 Remove about 10 cc. of heparinized blood by carotid cannulation.

2 Obtain plasma by centrifugation.

3 Prepare a smoked drum for the kymograph and set up the recording assembly, using an 8-inch straw for the recording arm. (To smoke drum, bubble illuminating gas through benzene.)

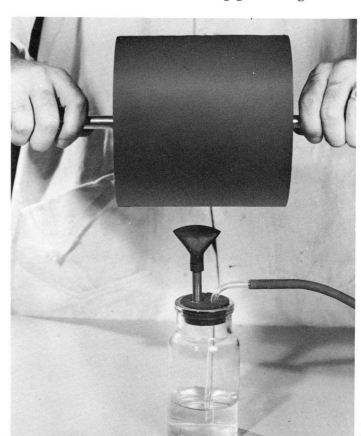

4 Fill the metal capsule with ethyl chloride. This is best accomplished by first cooling the apparatus used to handle the ethyl chloride in an ice bath. Place about 2 cc. of ethyl chloride in a crucible, then transfer the liquid to the metal capsule with a syringe. Fill the capsule about half full. Attach the capsule to the tambour by means of a rubber tube.

5 Place the capsule in a sample tube containing enough distilled water to cover it and place the sample tube in the ice-alcohol bath. Stir the ice mixture with the agitator and the water within the sample tube with the metal capsule assembly. As the water supercools, the writing needle will fall below the actual freezing-point of the water; but, as freezing takes place, the needle will rise and then level off. The rise is due to the heat given off during the freezing process (heat of fusion). The flat portion of the curve indicates the base line equivalent to 0° C.

6 Repeat the above procedure, using 1.7 per cent NaCl solution. This curve is equivalent to $-1°$ C.

7 Repeat the procedure, using clear plasma.

8 Remove the record from the kymograph drum and dip in shellac. Allow to dry.

CALCULATIONS

Draw horizontal lines through the flat portions of the curves representing the freezing-point in each case. The top line will represent 0° C., the bottom line $-1°$ C., and the middle line the relative freezing-point of the plasma. Measure the distance between the 0° line and the $-1°$ line and between the 0° line and the plasma line. The ratio of the two distances gives the freezing-point of the plasma. The following results are from the record shown with this experiment.

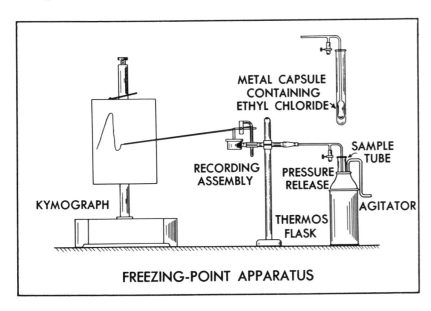

FREEZING-POINT APPARATUS

Record of freezing-point determination.

3 continued

	mm.
Distance between 0° line and −1° line	21.5
Distance between 0° line and plasma line	13.5
Distance between 0° line and 0.9 per cent NaCl line	11.5

$$\text{Freezing-point of plasma} = \frac{13.5}{21.5} = -0.63°\ \text{C.}$$

$$\text{Freezing-point of NaCl} = \frac{11.5}{21.5} = -0.54°\ \text{C.}$$

OSMOTIC-PRESSURE CALCULATION FROM FREEZING-POINT DATA

The osmotic pressure of blood can be calculated from the freezing-point of blood from the following expression.

$$\frac{\text{Freezing-point}}{1.87} \times 22.32 = \text{Osmotic pressure in atmospheres.}$$

$$\frac{0.63}{1.87} \times 22.32 = 7.5\ \text{atmospheres.}$$

The molal freezing-point constant = 1.87.

STUDENT'S REPORT

EXPERIMENT 4

EFFECTS OF CARBON DIOXIDE, LACTIC ACID, AND ASPHYXIA ON THE BLOOD PRESSURE

MATERIALS AND EQUIPMENT

Blood pressure setup.

Spirometer, as described in Part II, Experiment 2.

Tank of carbon dioxide;

    or the gas can be generated from sulfuric acid and calcium carbonate in a generator similar to that described in Part II, Experiment 2, omitting the NaOH absorption tube.

N/15 lactic acid (0.6 cc. conc. lactic acid plus 100 cc. water).

PROCEDURE

1. Cannulate the trachea.

2. Cannulate one of the carotids.

3. Cannulate the opposite femoral.

4. Prepare spirometer for attachment to the tracheal cannula.

5. Introduce into the spirometer enough carbon dioxide to give a concentration of 8–10 per cent.

6. Take a normal blood pressure tracing and then attach the tracheal cannula to the spirometer. In the rat the peripheral vasodilator effect usually predominates over the central constrictor effect, and a fall in the blood pressure results.

7. Now disconnect the spirometer and again obtain a normal record. Inject via the femoral vein 0.5 cc. of N/15 lactic acid solution. A prompt rise in blood pressure, marked especially by a strengthening of the pulse beat, occurs.

8. When the animal has returned to normal, clamp off the tracheal cannula. An initial fall, followed by a considerable rise in blood pressure, follows.

RESULTS— SAMPLE RECORDS

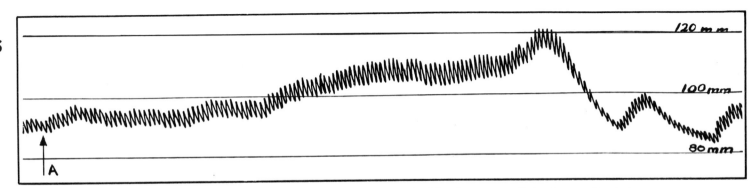

Effect of injection of 0.5 cc. of N/15 lactic acid on the blood pressure. Injection made at A.

4 continued

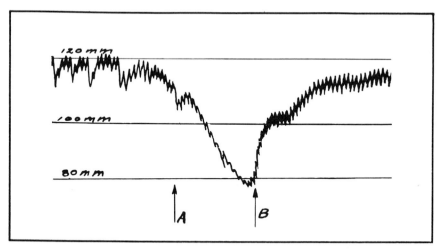

Effect of inhalation of approximately 10 per cent carbon dioxide on the blood pressure, at *A*; removed at *B*.

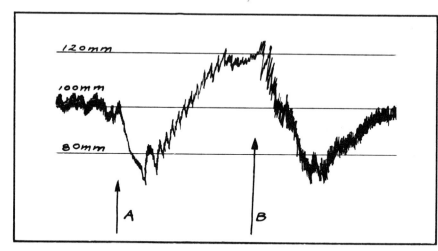

Effect of asphyxia; the trachea was clamped shut at *A* and released at *B*.

STUDENT'S REPORT

EXPERIMENT 5

EFFECT OF RUPTURE OF THE AORTIC VALVES ON THE BLOOD PRESSURE (WATER-HAMMER PULSE)

Any abnormality of the arterial system which permits blood to escape into a region of lower pressure, such as regurgitation into the left ventricle through the aortic valve, results in an increased pulse pressure. In this experiment the aortic valves are ruptured to produce this effect.

MATERIALS AND EQUIPMENT

Blood pressure setup.
Stilette with blunt, smoothly rounded tip.

PROCEDURE

1 Cannulate the left carotid for attachment to tambour.

2 Tie off the right carotid as far headward as possible, place a bulldog on the body end, and make a snip as for cannulation. Place a loop of thread around the right carotid, between the bulldog and the snip, in such a manner that pulling on its ends will occlude the vessel.

3 Attach the cannula to the tambour and take a normal tracing.

4 See that the tip of the stilette, which is to be passed downward through the carotid into the heart, is perfectly round and smooth. Insert the stilette into the cut made in the right carotid and pass it into the loop. Draw tightly on the loop and remove the bulldog. Loosen the loop a little and gently slide the stilette down the carotid toward the heart, exerting, meanwhile, suffi-

cient traction on the loop to prevent the loss of blood. The tip of the stilette is passed through the innominate artery, the root of the aorta, and into the ventricle. Resistance is felt when the valves are encountered, and they are ruptured by forcing the stilette through them. It is possible to pass the stilette into the ventricle without rupturing any valve; the stilette should be marked in advance, having been measured off on a previous carcass, to indicate the approximate level at which the tip reaches the valves. It may be necessary to withdraw the stilette

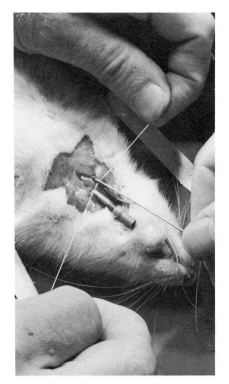
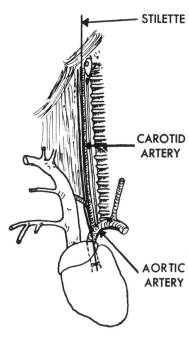

5 continued

a short distance and try again, if the first passage is unsuccessful. The arterial pulse can be felt along the stilette, and when a valve is ruptured a stronger pulse is perceived. As soon as the valves are ruptured, withdraw the stilette and tie the loop which held it in place. The pulse pressure will be found to be considerably increased above the normal.

RESULTS—SAMPLE RECORDS

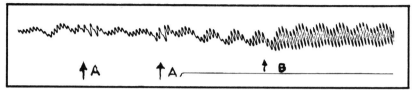

Stilette passed through the aortic valves at points marked *A* and withdrawn from the aorta at *B*.

STUDENT'S REPORT

EXPERIMENT 6

LATERAL VERSUS THRUST BLOOD PRESSURE

Sphygmomanometer determinations of blood pressure measure its lateral component; cannulation of an artery measures the total or thrust pressure. The difference between the two can be determined by using a **T**-shaped cannula, which also makes it possible to determine the back pressure from the circle of Willis.

MATERIALS AND EQUIPMENT

Blood pressure setup.

T-cannula.

PROCEDURE

1. In cannulating with the **T**-cannula a somewhat different procedure is adopted. Free as much of the carotid as possible and tie it off at the mid-point. Do not cut off the ends of the thread. Place a bulldog as far headward as possible, cut a snip, and insert one end of the crossbar of the cannula. Tie tightly, then cut the carotid completely through on the cannulated side of the central tie. Place another bulldog as far bodyward as possible and, by means of a bulldog or other weight attached to the thread of the central tie, exert slight traction. Lay the cannula alongside the free end of the artery to find the best place for the snip, make it, and, by means of the weight, maneuver the artery into the proper position for inserting the other end of the crossbar. Attach the syringe containing heparin to the central arm of the **T**, remove both bulldogs, and slowly inject the heparin. Have a short piece of rubber tubing attached to the central bar and clamp it off after the heparin has been given.

2. Now attach the central bar to the tambour. The record obtained is of the lateral pressure, since blood is flowing freely through the crossbar.

3. Place a bulldog on the head end of the artery; the pressure now obtained is the thrust pressure, which will be found to be from 10–20 mm. higher than the lateral pressure.

4. Remove the bulldog and now place it on the body end of the artery. The pressure now registered is the back pressure from the circle of Willis, which will be found to fall to 80 mm. or less.

6 continued

RESULTS—SAMPLE RECORDS

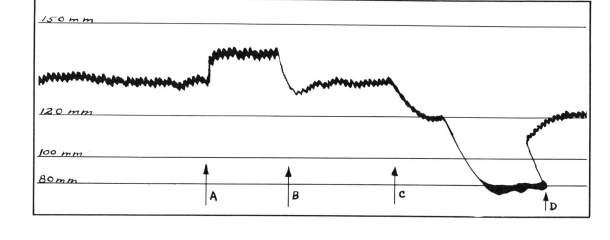

Portion of graph preceding A shows lateral blood pressure. The head end is occluded at A, giving thrust pressure; released at B. The body end is occluded at C; the pressure now registered is the back pressure from the circle of Willis.

STUDENT'S REPORT

EXPERIMENT 7

BLOOD VELOCITY

In this experiment the velocity of the blood in the carotid artery is determined by permitting it to flow through a measuring device called a "stromuhr."

MATERIALS AND EQUIPMENT

Usual surgical instruments.

Stromuhr.

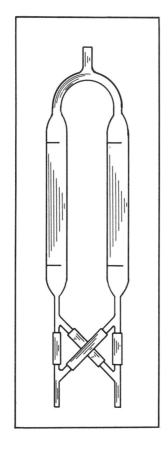

PROCEDURE

1. To prepare the stromuhr for use, fill one bulb with kerosene, the other with saline. This is easiest done by inverting the stromuhr, introducing enough saline to half-fill both bulbs, then, by means of a syringe, force air into one outlet until the saline fills one bulb; clamp off that outlet and fill the other bulb with kerosene by means of a long hypodermic needle and syringe. Clamp off both outlets and set up the stromuhr in the proper position to be placed over the rat.

2. Expose as long a section of the carotid as possible and tie it in the center. Cannulate the body end first and inject heparin, then cannulate the head end. (The method described in Part II, Exper. 6, with the T-cannula, is useful here.) For these cannulations, use the special curved cannulae which fit the stromuhr connections. Insert the cannulae vertically, so that the artery will not be twisted when the connections are made.

3. When the cannulations have been made, connect the cannulae to the stromuhr, with the body end of the carotid attached to the bulb containing kerosene and the head end to the saline bulb. Remove bulldogs and clamps. Upon entrance of the blood the kerosene is forced into the other side; the saline goes into the artery. (By means of a hot towel, the saline in the bulb has been kept warm.)

4. As soon as the blood reaches the lower mark, start the stop watch; when it reaches the upper, quickly change position of the clamps so that the direction of the blood flow is reversed.

7 continued

The blood flows back into the artery, and the other side fills. Repeat this several times, keeping count of the number of fillings and of the time required.

5   Measure as carefully as possible the internal diameter of the cannula. From the number of fillings and the time, the average time required to fill the calibrated portion of the bulb can be calculated, and when the diameter of the cannula is known, the velocity may be determined.

CALCULATIONS

Volume of bulb: 0.72 cc., or 720 cu. mm.
Average time of filling: 15 sec.
Radius of cannula: 0.3 mm. $\pi r^2 = 0.2826$.

$$\frac{720}{15} = 48 \text{ cu. mm/sec.}$$

$$\frac{48}{0.2826} = 170 \text{ mm/sec.}$$

STUDENT'S REPORT

EXPERIMENT 8

CIRCULATION TIME

In the first part of this experiment the time required for the blood to traverse the complete circulation, from femoral vein to the capillaries of the hind feet, is determined; in the second, the passage through the pulmonary circuit is timed.

MATERIALS AND EQUIPMENT

Purple X light bulb (General Electric Co.).
20 per cent sodium fluorescein solution.
1 per cent methylene blue solution.
0.1 per cent sodium cyanide solution.
    KCN is equally satisfactory.
Stop watch.
Ink-writing instrument.
Timer.
Signal magnet.

PART 1

When a solution of sodium fluorescein is injected and reaches the capillaries, the bare-skin regions (nose, ears, pads of feet, etc.) will fluoresce. This fluorescence is dramatically visible under the light emitted by a Purple X bulb. The observation is best made in a darkroom, although a box equipped with a peephole, inverted over rat and bulb, works quite well.

PROCEDURE

1 Anesthetize a rat and make a skin incision along the inner surface of the thigh, running from knee to body, thereby exposing the femoral vein. Separate it carefully from the femoral artery, along which it lies. (Refer to the diagram in Part I, Exper. 2, showing femoral cannulation.)

2 Move the animal to the darkroom. Using a short, 26-gauge needle, inject quickly, bodyward, 0.1 cc. of the fluorescein solution, giving a signal to the stop-watch operator at the moment of beginning the injection. Quickly switch the regular light off and the Purple X bulb on (or place the box quickly over the rat and bulb) and watch for fluorescence in the pads of the hind feet. Its appearance is sudden and sharp. Stop the

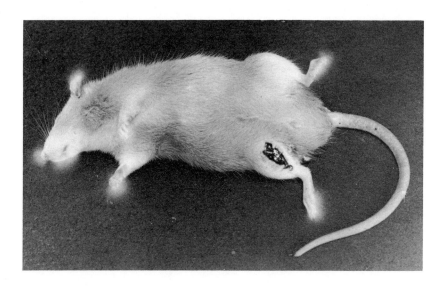

8 continued

watch the instant it is first noted. The time elapsed is the total circulation time and will be from 6 to 10 seconds. Control the hemorrhage from the vein with a bulldog or cotton pads; the animal may then be used for the remainder of the experiment.

PART 2

1 Now expose a carotid artery for as great a length as possible and prepare the jugular vein on the opposite side for intravenous injection. Place a small, trough-shaped piece of white cardboard or celluloid beneath the artery, being careful that its edges do not obstruct the blood flow and make sure of adequate illumination.

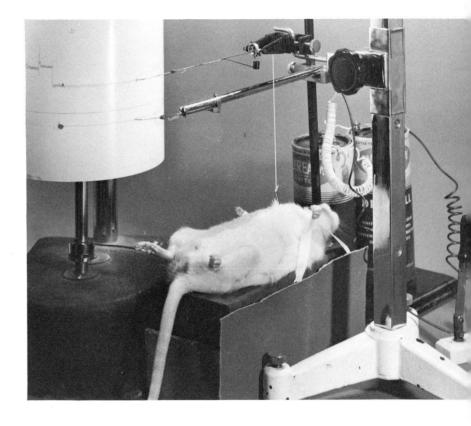

2 Inject quickly 0.5 cc. of the methylene blue solution into the jugular, giving a signal to the stop-watch operator at the instant of starting the injection. Watch for the appearance of the methylene blue in the blood passing through the carotid. The time elapsed is the pulmonary circulation time and is found to be about 3 seconds.

Another method of determining the pulmonary circulation time makes use of the quickening of respiration produced by an injection of sodium cyanide. This drug acts upon the aortic and carotid bodies; when injected into the jugular vein the pulmonary circulation is traversed before these bodies are reached.

3 Having marked 1-second time intervals on a fairly fast kymograph drum, arrange for a respiration record by attaching a bent pin to the thorax and fastening it by means of a thread to the ink-writing instrument. Adjust the signal magnet to write directly below the respiration record.

4 Now inject via the jugular vein 0.05 cc. of the 0.1 per cent sodium cyanide solution, the switch of the signal magnet circuit being closed at the same instant. Within a few seconds an increase in the rate of respiration will be recorded.

RESULTS—SAMPLE RECORDS

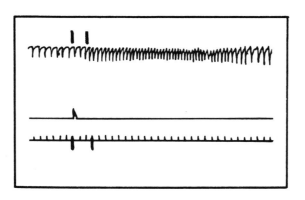

Graph illustrating cyanide method for determining pulmonary circulation time. Injection made at point marked by signal magnet (*middle line*). Pulmonary circulation time is 3 seconds. Note great increase in respiration rate produced by cyanide.

STUDENT'S REPORT

EXPERIMENT 9

DETERMINATION OF THE CARDIAC OUTPUT

The cardiac output can be calculated from the oxygen consumption and the arteriovenous oxygen difference by application of the Fick principle. The oxygen consumption is determined by direct measurement with a recording microspirometer; arterial blood samples are obtained by cannulation of the carotid artery, and mixed venous blood is obtained by introducing a cardiac cannula directly into the heart. The blood samples are analyzed for their oxygen content, using the micromethod of Scholander.

The cardiac output is influenced by a number of factors, including drugs such as adrenalin, acetylcholine, nitrites, etc. In this experiment the effect of adrenalin on the cardiac output is demonstrated.

MATERIALS AND EQUIPMENT

Instruments for cannulation.
Heart cannula.
> 18-gauge needle about 4 cm. long, which has previously been marked at the proper depth of introduction, using a rat cadaver.

Oxygen consumption apparatus.
Kymograph.
Stop watch.
Scholander Roughton analyzer (may be obtained from A. S. Aloe Co., St. Louis, Mo.)
Caprylic alcohol.

Solutions for Scholander method.
> Ferricyanide solution: 6.3 gm. $K_3Fe(CN)_6$, 1.5 gm. $KHCO_3$, and 0.25 gm. saponin dissolved in 25 cc. water.
> Acetate buffer: 17.5 gm. $NaC_2H_3O_2 \cdot 3H_2O$ per 25 cc. of water, add 3.75 cc. glacial acetic acid.
> 45 per cent urea solution.
> 10 per cent NaOH.
> Pyrogallol solution: 3.75 gm. pyrogallol per 25 cc. 20 per cent NaOH. Protect from air with a layer of mineral oil.
> The solutions used for the Scholander procedure are most easily added from 5-cc. syringes.

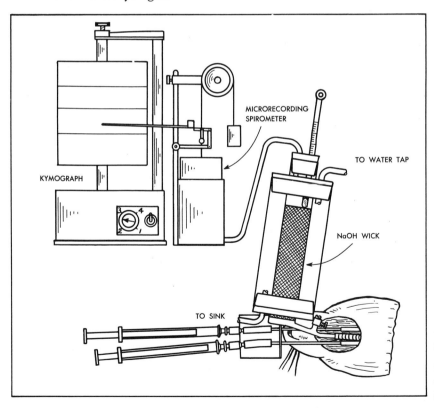

PROCEDURE

Set up oxygen consumption apparatus.

1. Soak wick in 20 per cent NaOH and insert in the absorption tube. Fill the water jacket with water at room temperature or just below.

2. With the absorption tube attached to the microspirometer, calibrate the apparatus by drawing lines around the kymograph drum at 5-cc. intervals.

3. Fill the microspirometer and absorption tube with oxygen.

4. Cannulate the trachea and left carotid artery; inject 0.25 cc. heparin.

5. Introduce the cardiac cannula into the right external jugular vein, through the superior vena cava and into the right auricle. The end of the cannula should move up and down with each heart beat.

6. Connect the arterial and cardiac cannulae to the tubes mounted in the support, which is clamped to the animal board. Connect 1-cc. syringes to each of the tubes.

7. Attach the rubber tube from the absorption apparatus to the trachea of the animal. The writing needle will rise and fall with each respiration but will also show a fall due to oxygen consumption. When the needle just crosses one of the calibration lines, start the stop watch. Then withdraw samples of arterial and mixed venous blood while oxygen consumption is being determined. After the samples have been obtained, record the time and temperature just as the needle crosses one of the other calibration lines.

8. The blood samples can be stored anaerobically in the 1-cc. syringes by plugging the tips of the syringes with round toothpicks. If the analysis for blood oxygen is not to be completed immediately, the samples should be stored in the refrigerator.

9. Inject intravenously 0.005 mg/kg of adrenalin and immediately determine the cardiac output as above.

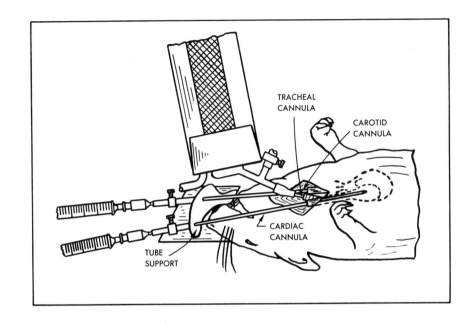

ANALYSIS OF BLOOD SAMPLES

1. Mount the Scholander syringe vertically and rinse several times with ferricyanide solution. Care must be taken to prevent trapping of air.

2. Fill the glass cup to the mark with ferricyanide and draw down to the bottom of the cup.

3. Place a drop of caprylic alcohol in the bottom of the cup.

4. Fill the micropipette to the mark with blood and hold at a slight angle to the horizontal, so that the blood does not flow out when both ends of the pipette are open to the air. With

9 continued

the Scholander syringe at the same angle, the pipette is cautiously introduced into the glass cup. Press the tip snugly against the bottom of the cup.

5. By pulling the plunger gradually, the blood is slowly drawn into the capillary, followed by a bubble of air about 1 mm. in length.

6. Remove the pipette and expel the air through the caprylic alcohol with the aid of a fine wire.

7. Draw a trace of caprylic alcohol into the capillary and remove the remainder from the cup.

8. Fill the cup to the mark with acetate buffer and draw the buffer to the bottom of the cup.

9. Fill the cup with urea solution and close firmly with the finger.

10. Shake the closed apparatus in the horizontal position for about 2 minutes; draw the plunger out gradually as the gases are evolved, so as to maintain the gas pressure in the syringe at about atmospheric pressure.

11. Release the finger cautiously, with manipulation of the plunger so as to keep the gas meniscus in the capillary. Allow a small amount of urea solution to pass down the capillary to clean the walls.

12. Three-quarters of the urea solution in the glass cup is removed, and the rubber cup is adjusted and filled with 10 per cent NaOH without trapping air bubbles.

13 A little NaOH is drawn into the syringe. This absorbs some CO_2, causing a partial vacuum, which quickly sucks in more NaOH until only a small bubble, consisting of O_2 and N_2, remains. Just before absorption is complete, transfer the bubble into the capillary by manipulation of the plunger.

14 Remove NaOH from the rubber and glass cups; then remove the rubber cup.

15 Place the capillary in a beaker of water at room temperature for ½ minute; remove and dry by light wiping.

16 Read the volume of the bubble. This is V_1.

17 Fill the glass tube with pyrogallol solution and absorb the O_2 by pulling the bubble down to the bottom of the capillary and back again several times. Adjust the temperature and read the volume. This is V_2.

18 Determine the blank by a similar experiment without added blood.

CALCULATIONS

1. Oxygen content of the blood equals $(V_1 - V_2 - c) \times f$,
   c = Blank,
   f = Correction for temperature, aqueous vapor pressure, and barometric press.
2. Cardiac output equals:
$$\frac{O_2 \text{ consumption (cc/min)}}{\text{Arteriovenous diff. (cc/100 ml blood)}} \times 100.$$

RESULT OF TYPICAL EXPERIMENT

Control.
- Arterial blood ... 25 cc. O_2/100 cc.
- Venous blood ... 17 cc. O_2/100 cc.
- A − V diff. ..... 8.0 cc.
- O_2 Cons. .... 3.5 cc/min.
- $\dfrac{3.5}{8.0} \times 100 = 44$ cc/min.

During 1 minute following injection of 0.005 mg/kg adrenalin.
- Arterial blood ... 24.2 cc. O_2/100 cc.
- Venous blood ... 21.5 cc. O_2/100 cc.
- A − V diff. ..... 2.7 cc.
- O_2 Cons. .... 3.8 cc./min.
- $\dfrac{3.8}{2.7} \times 100 = 140$ cc/min.

STUDENT'S REPORT

EXPERIMENT 10

PRODUCTION OF RENAL HYPERTENSION

This experiment introduces a method for the experimental production of hypertension by compression of the kidneys. This compression results in a reduction of blood flow, in consequence of which the kidney secretes a vasopressor agent known as "renin."

The so-called "tail method" of determining blood pressure is employed.

MATERIALS AND EQUIPMENT

Blood pressure apparatus.

This method requires a cuff with mercury manometer attached, a screw-capped gland, and a plethysmograph.

Packing cuff.

1. A 5-cm. length of drainage tubing having a diameter of 1 cm. is used. One end is attached to one arm of a T-tube, the other end is sealed with rubber cement.
2. Curve the sealed end upward and bind it to the arm of the T-tube by means of rubber stripping dipped in rubber cement. This forms an airtight cuff approximating the diameter of the average rat's tail.
3. In order to prevent outward expansion when pressure is applied, surround the exterior of the cuff with a strip of canvas, fastening its ends to the arm of the T-tube.
4. Attach a sphygmomanometer bulb to another arm of the T-tube.
5. Attach a mercury manometer to the remaining arm.

Packing gland.

1. A cylinder with one end closed and the other fitted with a screw cap is employed. The brass case supplied with a microscope objective is ideal for this purpose. Drill two holes (1.0 and 0.5 cm. in diameter) directly opposite each other perpendicular to the vertex of the cylinder at a point 2 cm. from its base. Pack the cylinder with green soft soap or silicone stopcock grease.

Plethysmograph.

1. The plethysmograph is a glass tube about 12 cm. in length and 1 cm. in inside diameter. It bears a capillary tube of ½-mm. bore and a side tube for filling and adjusting the level of the fluid in the capillary tube. One end of the plethysmograph is to be attached to the gland; a thermometer is inserted into the other.
2. The temperature of the water in the plethysmograph should be maintained at about 40° C. To accomplish this, wrap a small-gauge, copper wire spirally around the tube, attaching its ends to a transformer. The voltage required will depend on the length and gauge of the wire and will have to be determined by trial.

Sterile gauze.

Cellulose acetate.

PROCEDURE

Young (100–125 gm.) rats are used. Since the blood pressure determination takes only a few minutes, tame rats, held lightly covered with a towel, usually remain quiet during the test; or they may be lightly anesthetized with Nembutal (25 mg/kg). Several determinations are made on each rat, and the results averaged. They are then rendered hypertensive, as described below, and the blood pressure determinations repeated some 30 days later.

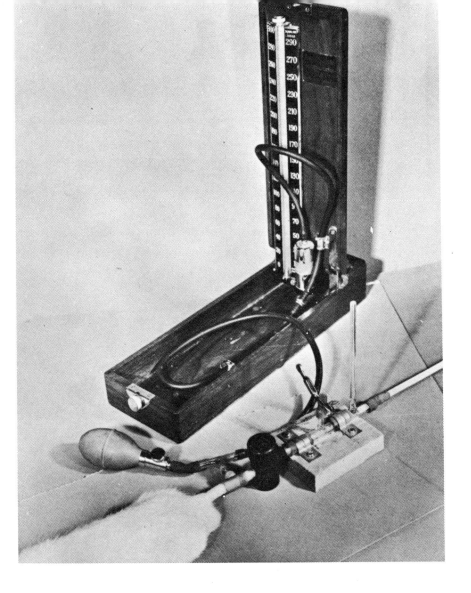

The blood pressure is determined as follows:

1. Place the cuff around the rat's tail 1 cm. distal to the root.

2. Remove the cap of the metal gland, pack it with soft soap, and replace the cap with the thread lightly engaged. Remove the central core of soap, i.e., that lying between the two drilled holes. Attach the plethysmograph to the gland by means of a short rubber tube, one end of which fits over the plethysmograph tube and the other inserts snugly into the large drilled hole of the gland.

3. Insert the rat's tail into the plethysmograph through the small drilled hole to within 2 cm. of the cuff. Screw the cap of the gland down until the tail is engaged by the soap.

4. Fill the plethysmograph with water at about 40° C. and adjust the level in the capillary to a convenient point.

5. Arrest the circulation by rapid inflation of the cuff to a pressure above the systolic level.

6. Allow the pressure to fall slowly by means of the release valve of the bulb, watching the level of fluid in the capillary tube meanwhile. When the systolic pressure is passed, an abrupt rise in the capillary tube will occur. The pressure may then be read from the manometer.

The method for producing renal hypertension is described below. However, this operation must be performed under sterile conditions, since the animals are to be kept alive for some time. Directions for the conduct of such experiments are given in Part I, Experiment 24.

continued

1. Anesthetize the rat with Nembutal.

2. Place the animal in a prone position and make a dorsal midline incision extending from about T10 to L3.

3. Deflect the skin to the sides, dissect through the thin muscle walls just anterior to the upper poles of the kidneys, lift the kidneys from their beds, and clear away the surrounding fat. The pad of fat at the upper pole, inclosing the adrenal gland, should be freed very carefully.

4. Encircle the kidneys with sterile gauze, applying the first layer parallel to the long axis of the kidney. Further layers are applied circularly, care being taken not to inclose the blood vessels.

5. Coat the gauze with several layers of thin cellulose acetate, permitting each coat to dry before applying the next. Drying is rapid, only a minute or two being required for each coat.

6. Replace the kidneys, close the muscle wound with sutures and the skin incision with wound clips.

TYPICAL RESULTS

From 10 animals

	Low	Average	High
Normal blood pressure	78 mm.	103 mm.	132 mm.
Blood pressure 30–45 days after encapsulation	153 mm.	175 mm.	188 mm.

STUDENT'S REPORT

EXPERIMENT II

ARTIFICIAL RESPIRATION BY THE DRINKER RESPIRATORY CHAMBER METHOD

In this experiment, complete respiratory paralysis is produced by Intocostrin, a curare preparation. Respiration is maintained by means of an apparatus similar to the Drinker respiratory chamber, its effectiveness being demonstrated by maintenance of the blood pressure and eventual recovery of the animal.

MATERIALS AND EQUIPMENT

Drinker respiratory chamber.

Blood pressure setup.

Valve and collecting-bulb system.

Intocostrin solution.

> This is Squibb's commercial preparation. It comes in 5-cc. ampoules, the potency being 20 units per cubic centimeter. The dose for the rat is 1.5 units per kilogram, which gives complete respiratory paralysis for from 15 to 20 minutes. It should be given intravenously in a volume of 0.2 cc. per 100 gm. of rat; therefore, dilute 1 cc. of Intocostrin with 25.6 cc. of saline.

PROCEDURE

1 Anesthetize a rat and cannulate the trachea.

2 Cannulate a carotid artery and administer heparin.

3 Cannulate the opposite jugular.

4 With the Drinker apparatus set up, lay the rat on the tray which fits into the jar and attach the tracheal cannula to the outlet tube and the carotid cannula to its proper connection in the lid.

5 Now inject, via the jugular, 0.2 cc. of the Intocostrin solution per 100 gm. of rat. Note that complete paralysis of the voluntary muscles occurs almost immediately, including the respiratory muscles.

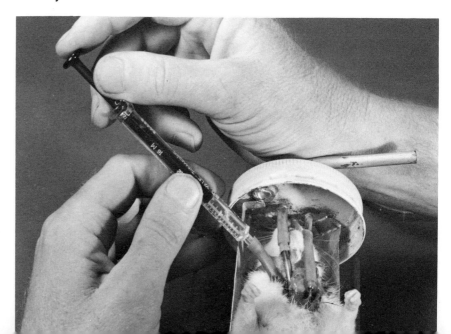

 continued

collapse when the pressure is increased. Positive and negative pressures of 10–20 mm. of mercury are adequate.

7 Attach the tracheal outlet to the valves and collecting bulbs and the carotid outlet to the tambour. The blood pressure record is the index to the rat's condition; should it tend to fall, the pressures being supplied to the chamber are probably inadequate. After 10 minutes, stop the artificial respiration momentarily to see whether the rat has begun to breathe. Repeat at intervals of a few minutes until it does. The pulmonary ventilation can be measured by means of the valves and collecting bulbs; it should be maintained at about 100 cc. per minute. It may be interesting to note how the pulmonary ventilation alters with changes in the pressures being introduced into the chamber.

6 At once slide the tray into the jar, screw the jar onto the lid, and start introducing alternately positive and negative pressure into the chamber by means of the syringe. Note the expansion of the rat's thorax as the pressure is reduced and its

RESULTS—SAMPLE RECORDS

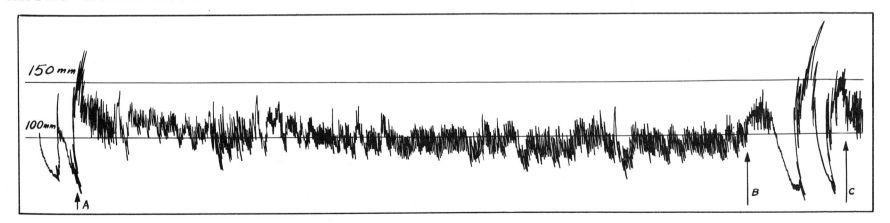

Record of blood pressure of rat with complete respiratory paralysis, respiration being maintained by the Drinker chamber.

At A artificial respiration was started; at B it was suspended temporarily; from C on, animal maintained its own respiration.

STUDENT'S REPORT

EXPERIMENT 12

DETERMINATION OF OXYGEN CONSUMPTION, RESPIRATORY QUOTIENT, AND BASAL METABOLIC RATE

In this experiment the amount of oxygen consumed and the amount of carbon dioxide produced, per minute of time, are determined. From these figures the respiratory quotient is calculated, and the basal metabolic rate, namely, the calories of heat produced per square meter of body surface per hour, is determined.

MATERIALS AND EQUIPMENT

1 N sodium hydroxide solution (4.0 gm. of NaOH per 100 cc. water).

30 per cent sodium hydrosulfite solution.

    Dissolve 15 gm. of sodium hydrosulfite in 50 cc. of 4 N potassium hydroxide. Filter quickly and store under oil until used (4 N KOH contains 22.4 gm. per 100 cc.).

Apparatus for collecting expired air, as described in Part I, Experiment 13.

Van Slyke manometric apparatus.

50-cc. syringe.

PROCEDURE

1 Cannulate the trachea and attach valve system to trachea and collecting bulb. The collecting bulb should be equipped with a threeway stopcock.

2 Collect about 100 cc. of air, discard through the other outlet of stopcock. This is for the purpose of washing atmospheric air out of the tubes.

3 Now time accurately the collection of 100–150 cc. of air. This is the air used for analysis and must consist of only expired air; it is therefore important to maintain a slight positive pressure against the valves to prevent leakage.

4 Attach a short rubber tube to one of the stopcock outlets and fill it and the outlet with light mineral oil. The plunger of the 50-cc. syringe should also be oiled. Attach the syringe to the rubber tube and withdraw about 50 cc. of the expired air.

5 Remove the rubber tube along with the syringe, and to it attach a short piece of capillary tubing the end of which is fitted with a rubber collar. The syringe will contain some oil which can be used to fill the capillary tube. The sample is now ready for introduction into the Van Slyke.

6 With the bulb in position *1*, permit the mercury to flow slowly upward through the upper stopcock and into the chamber until the chamber contains 1 or 2 cc.

7 Dry the surface of the mercury with a filter paper. Leaving some mercury in the chamber and stopcock to serve as a seal, draw the remainder down to slightly below the 50–cc. mark below the bulb.

8 Place leveling bulb in position *2* and bring the mercury exactly to the 50-cc. mark. Read the pressure, P_0.

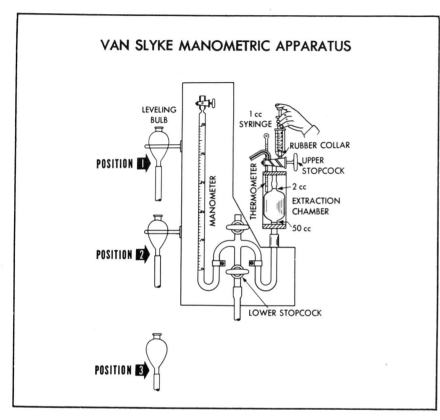

VAN SLYKE MANOMETRIC APPARATUS

9 With leveling bulb in position 2 and lower stopcock open, introduce about 35 cc. of the expired air sample by holding the rubber-tipped end of the capillary tube tightly against the flare of the chamber above the stopcock. The excess oil is removed by exerting pressure on the plunger and forcing it out below the mercury before opening the stopcock. Open the stopcock and let the air enter.

10 Again take the mercury down to the 50-cc. mark and read the pressure, P_1. Read the temperature.

11 Introduce by means of a 1-cc. syringe, the tip of which is equipped with a rubber collar, exactly 1 cc. of 1 N sodium hydroxide solution. Raise the mercury until it fills approximately the lower third of the bulb, and shake for 2 minutes.

12 Lower below the 50-cc. mark and bring the mercury meniscus (not the solution) exactly to the 50-cc. mark and read the pressure, P_2.

13 Admit 3 cc. of the 30 per cent hydrosulfite solution in the same manner, shake for 3 minutes with the mercury about one-third up the bulb, bring down to the 50-cc. mark, and, with the mercury meniscus on the mark, read the pressure, P_3.

14 Now eject the remaining gases from the chamber without loss of solution, bring the mercury meniscus down to the 50-cc. mark, and read the pressure, P_4.

15 Eject the solution and clean the apparatus as described in Part II, Experiment 1.

CALCULATIONS

Refer to Peters and Van Slyke, *Quantitative Clinical Chemistry*, 2:114:

Weight of rat: 330 gm.

Expired air: 110 cc/min.

Temp. 26°5

P_0, 227 mm.

P_1, 630 mm.

P_2, 615 mm.

P_3, 574.5 mm.

P_4, 226.5 mm.

P_1 minus P_0 = 403 mm., pressure of sample.

(P_2 minus P_0) 0.98 = 380.2 mm., pressure of oxygen and nitrogen.

Pressure of carbon dioxide = 403 − 380.2 = 22.8 mm.

(P_3 minus P_4) 0.92 = 320.2 mm., pressure of N_2.

Pressure of oxygen = 380.2 − 320.2 = 60 mm.

12 continued

$$\text{Percentage carbon dioxide} = \frac{22.8}{403} \times 100 = 5.6 \text{ per cent.}$$

$$\text{Percentage oxygen} = \frac{60}{403} \times 100 = 14.9 \text{ per cent.}$$

$$\text{Respiratory quotient} = \frac{\text{Vol. carbon dioxide produced}}{\text{Vol. oxygen consumed}} = \frac{5.6}{21 - 14.9} = 0.91.$$

Oxygen consumption: 110 cc. expired air $\times$ 6.1 per cent = 6.7 cc. O_2/min.

$$6.7 \times \frac{630^*}{760} \times \frac{273}{299.5} = 5.0 \text{ cc. } O_2/\text{min at standard conditions.}$$

* Barometric pressure in Denver is 630 mm.

At an R.Q. of 0.91, 1 liter oxygen = 4.93 cal., $5.0 \times 60 = 300$ cc/hr, or $0.3 \times 4.93 = 1.48$ cal/hr.

A formula (Hill and Hill) used for calculating the surface area of the rat is:

$$(\text{Weight in gm.})^{2/3} \times 10 = \text{Surface area in square centimeters.}$$

This rat weighed 330 gm.

$$(330)^{2/3} \times 10 = 470 \text{ sq. cm.} = 0.047 \text{ sq. m.}$$

$$\frac{1.48}{0.047} = 33.6 \text{ cal/sq m/hr.}$$

STUDENT'S REPORT

EXPERIMENT 13

THE BROMSULPHALEIN TEST FOR LIVER FUNCTION

One test for liver function is the rate of removal of the dye, bromsulphalein, from the blood. In this experiment the retardation of removal is demonstrated when liver function is impaired.

MATERIALS AND EQUIPMENT

Instruments for cannulation
Trachea, carotid, and femoral cannulae.
Cannula for bile duct (as shown in Part I, Exper. 21).
Bromsulphalein (BSP) solutions.
    Concentrations of 2 mg/cc and 0.02 mg/cc (stock solution containing 50 mg/cc can be obtained from Hynson, Westcott and Dunning).

0.1 N NaOH.
Stop watch.
Colorimeter.
Centrifuge.

PROCEDURE

1. Cannulate the trachea, left carotid artery, and right femoral vein. Heparinize.

2. Cannulate the bile duct as described in Part I, Exper. 21.

3. Remove 0.5 cc. of blood from the carotid artery; transfer to a 15-cc. centrifuge tube.

4. Inject BSP via the femoral vein (5 mg/kg); start the stop watch. Place a small dish containing a few cc. of 0.1 N NaOH under the end of the bile cannula. The BSP is removed from the blood stream by the liver and excreted in the bile. When the bile containing the BSP comes in contact with the alkali, it turns it purple; the first dye appears in 3–4 minutes.

5. Two minutes after the injection of BSP, collect another 0.5-cc. sample of blood and transfer to the second centrifuge tube.

6. After about 30 minutes, during which time most of the dye has been removed from the blood, clamp off two lobes of the liver with hemostats so as to reduce the effectiveness of the liver.

7. Repeat the injection of BSP as in part 4.

8. Collect a 0.5-cc. sample of blood, 2 minutes after the BSP injection, and transfer to a third centrifuge tube.

9. Centrifuge the three blood samples, measure 0.2 cc. of plasma from each, and transfer to test tubes.

10. To the first test tube add 2.0 cc. of H_2O, 0.5 cc. of the BSP solution containing 0.02 mg/cc and 3.0 cc. of 0.1 N NaOH. To tubes 2 and 3 add 2.5 cc. of H_2O and 3 cc. of 0.1 N NaOH.

13 continued

11 Compare the resulting solutions in a colorimeter.

12 Calculate the per cent retention of the dye in an animal with normal liver function and when a portion of the liver was clamped off; assume the plasma volume to be 3 per cent of the body weight.

STUDENT'S REPORT

CALCULATIONS

Weight of rat: 200 gm.

Calculated plasma volume: $200 \times 0.03 = 6.0$ cc.

$Rs = 20.0$ mm.

Rx (first sample) $= 30$ mm.

Rx (second sample) $= 10$ mm.

$Cs = 0.01$ mg. B.S.P.

$\dfrac{20}{30} \times 0.01 \times \dfrac{6}{0.2} = 0.2$ mg. B.S.P. remaining after 2 minutes in the normal animal.

$\dfrac{0.2}{1.0} \times 100 = 20$ per cent retention.

$\dfrac{20}{10} \times 0.01 \times \dfrac{6}{0.2} = 0.6$ mg. B.S.P. remaining.

$\dfrac{0.6}{1.0} \times 100 = 60$ per cent retention.

EXPERIMENT 14

THE RATE OF GLUCOSE ABSORPTION (CORI AND CORI COEFFICIENT)

In this experiment the rate of absorption of glucose, expressed in terms of the Cori and Cori coefficient, namely, the number of milligrams per 100 gm. of rat per hour, is determined.

MATERIALS AND EQUIPMENT

Syringe and needle for oral administration.

Reagents for Schaffer-Somogyi method for sugar determination.

PREPARATION OF REAGENTS

Copper reagent.

	Gm.
Sodium carbonate (anhydrous) (Na_2CO_3)	25
Sodium bicarbonate ($NaHCO_3$)	20
Rochelle salt	25
Copper sulfate ($CuSO_4 \cdot 5H_2O$)	7.5
Potassium iodate (KIO_3)	1.07
Potassium iodide (KI)	1.0

Dissolve the $NaHCO_3$ and Rochelle salt in 600 cc. of distilled water. Add the $CuSO_4 \cdot 5H_2O$ (dissolved in 75 cc. of water) slowly through a funnel, stirring as it is added. Add the $NaHCO_3$ and KI and stir until dissolved. Rinse into a liter volumetric flask. Add the KIO_3, dilute to volume, and filter. Protect from the sunlight.

0.1 N sodium thiosulfate.

Add 25 gm. $Na_2S_2O_3 \cdot 5H_2O$ to about 100 cc. of water and dissolve. Make up to a volume of 1 liter. Standardize against 0.1 N KIO_3 (3.567 gm. per liter), using 20-cc. portions with 1.5–2.0 gm. of KI and 15 cc. of 1 N sulfuric acid. Use starch as an indicator.

$KI-K_2C_2O_4$ solution.

Prepare solution containing 2.5 per cent of KI and potassium oxalate ($K_2C_2O_4$) in a dark bottle. (Is stable for one week only.)

N sulfuric acid.

Prepare a N sulfuric acid (26 cc. of concentrated acid diluted to 1 liter).

Starch solution.

Prepare a starch solution (1 per cent solution of soluble starch).

PROCEDURE

1 Administer orally 5 gm. of glucose per kilogram body weight, in a 50 per cent solution, to a rat which has been starved for 24 hours.

2 After 2 hours, kill the rat, tie off the esophagus at the cardiac sphincter and the small intestine at the iliocolic sphincter, and remove the stomach and small intestine. Wash the blood from the outside. Slit the tract open and wash out the content thoroughly with water.

3 Filter if necessary, dilute to 400 cc., and measure 5-cc. aliquots into large Pyrex test tubes. Measure 5 cc. of water into a test tube to be treated in the same manner as the unknowns.

4 Add 5 cc. of the copper reagent to each tube, washing down the sides; mix the samples and cover with one-hole rubber stoppers to avoid oxidation.

5 Place in a boiling water bath for 15 minutes with as little agitation as possible.

14 continued

STUDENT'S REPORT

6 Cool in running water and add 2 cc. of the KI–$K_2C_2O_4$ solution and 5 cc. of N sulfuric acid. Cover with a small beaker and agitate.

7 After a few minutes wash the beakers into the test tubes and titrate with 0.005 N sodium thiosulfate solution, made by diluting the 0.1 N. When the brown iodine color becomes yellow, add 2–3 drops of the starch solution and continue the titration until the starch blue is gone and the delicate copper green is visible.

CALCULATIONS

Subtract the cubic centimeters of thiosulfate required to titrate the unknown sample from the amount required to titrate the blank run with water. Multiply this figure by 0.112, which is an average figure for solutions containing 0.1–2.5 mg. of glucose per 5 cc., as determined experimentally with 0.005 sodium thiosulfate. This figure represents the amount of glucose in 5 cc. and must be multiplied by 80 to give the total amount. This amount, subtracted from the glucose administered, gives the amount absorbed.

RESULTS OF TYPICAL EXPERIMENT

Weight of rat: 250 gm.
Amount of glucose administered: 1,250 mg.
Blank: 28.1 cc.
Unknown: 4.1 cc.
Difference: 24.0 cc. $\times$ 0.112 = 2.7 mg/5 cc.

$2.7 \times \dfrac{400}{5} = 216$ mg., amount unabsorbed.

$1{,}250 - 216 = 1{,}034$ mg., absorbed in 2 hours = 517 mg. per hour.

$517 \times \dfrac{100}{250} = 207$ mg., absorbed per hour per 100 gm. of rat.

(The Coris found the range of coefficient to lie between 180 and 220.)

EXPERIMENT 15

EFFECT OF ADRENALIN ON BLOOD SUGAR

MATERIALS AND EQUIPMENT

Adrenalin solution, containing 0.01 mg/cc.

Reagents for Folin-Malmros blood sugar method.

PREPARATION OF REAGENTS

Dilute tungstic acid solution.

Transfer 20 cc. of 10 per cent sodium tungstate to a 1-liter volumetric flask. Dilute to about 800 cc. Add, with shaking, 20 cc. of a ⅔ N sulfuric acid solution and dilute to volume.

Sodium cyanide–carbonate solution.

Transfer 8 gm. of anhydrous sodium carbonate to a 500-cc. volumetric flask. Add 40–50 cc. of water and shake. Add 150 cc. of freshly prepared 1 per cent sodium cyanide solution, dilute to volume, and mix.

Potassium ferricyanide solution.

Dissolve 2 gm. of c.p. potassium ferricyanide in distilled water and dilute to a volume of 500 cc. Keep major part of solution in a brown bottle.

Ferric iron solution.

Fill a liter cylinder with distilled water. Suspend, on a copper-wire screen, just below the surface, 20 gm. of soluble gum ghatti, and leave overnight. Remove the screen and strain the liquid through a double layer of clean towel. Add to this extract a solution of 5 gm. of anhydrous ferric sulfate in 75 cc. of 85 per cent phosphoric acid plus 100 cc. of water. Add to the mixture, a little at a time, about 15 cc. of a 1 per cent potassium permanganate solution to destroy reducing materials present in gum ghatti. The slight turbidity of the solution will disappear in a few days.

Standard glucose solution.

A stock solution containing 10 mg. of glucose per cubic centimeter is prepared in saturated benzoic acid by dissolving 6 gm. of benzoic acid in 100 cc. of boiling water, cooling somewhat, filtering, and adding 1 gm. of glucose. From the stock solution, a solution containing 2 mg. per cubic centimeter is prepared, which may be further diluted according to the value of the sample.

PROCEDURE

1 Anesthetize a rat with Nembutal, but do not use ether.

2 Cannulate a carotid and administer heparin; in injecting solutions alternately withdraw and reinject blood several times.

3 Take a sample of blood (not more than 0.5 cc.) for the normal determination.

4 Inject, very slowly, 0.01 mg. adrenalin per kilogram of rat; as soon as the rat has recovered, as indicated by a return to normal respiration, repeat the injection.

5 Wait 30 minutes, keeping animal warm and the wound moist with warm saline packs, and then withdraw another blood sample.

6 With an accurate 0.1-cc. pipette, measure 0.1 cc. of each blood sample and transfer to 10 cc. of dilute tungstic acid solution in a small vessel. Mix well and filter.

7 Transfer 4 cc. of the water-clear filtrate into a Folin sugar tube. Transfer 4 cc. of the standard sugar solution into a similar

15 continued

tube. (For the normal, use a standard containing 0.01 mg. glucose per cubic centimeter; for the second sample use a standard containing 0.02 mg. per cubic centimeter.) To each tube add 2 cc. of the potassium ferricyanide solution and 1 cc. of the cyanide-carbonate solution.

8 Heat in boiling water for 8 minutes and cool in running water for 1–2 minutes.

9 Add 5 cc. of the ferric iron solution and mix.

10 Let stand for 1–2 minutes and then dilute with water almost to the 25-cc. mark. Add 2 drops of alcohol to cut the foam and dilute to the mark. Mix.

11 Read in the colorimeter fitted with a picric acid light filter. (To prepare the light filters, soak 8–10 sheets of filter paper with a solution of 5 gm. of picric acid dissolved in 100 cc. methyl alcohol, to which 5 cc. of a 10 per cent sodium hydroxide solution has been added. Use the dry papers to cover the light box of the colorimeter.)

RESULTS OF TYPICAL EXPERIMENT

Normal: 0.01 mg/cc used as standard; standard reading: 20; unknown reading: 16.

$$\frac{0.01 \times 20}{16} = 0.0125 \text{ per } 0.01 \text{ cc. of blood} = 125 \text{ mg}/100 \text{ cc.}$$

After adrenalin: 0.02 mg/cc standard; standard reading: 20; unknown reading: 19.

$$\frac{0.02 \times 20}{19} = 0.021 \text{ mg}/0.01 \text{ cc of blood} = 210 \text{ mg}/100 \text{ cc.}$$

STUDENT'S REPORT

EXPERIMENT 16

EFFECT OF INSULIN ON BLOOD SUGAR

MATERIALS AND EQUIPMENT

Insulin solution containing 2 units per cubic centimeter.
>The regular, not modified, insulin should be used.

Reagents for Folin-Malmros sugar-determination method, as described in the previous experiment.

PROCEDURE

1. The animals used should be starved for 48 hours.

2. Anesthetize lightly with Nembutal.

3. Take a sample of tail blood for the normal determination.

4. Inject 2 units of insulin per kilogram, intraperitoneally.

5. After 30 minutes, take samples of tail blood. The rats may or may not be in convulsions, but have at hand 5 per cent glucose solution to relieve the hypoglycemic shock after the samples have been taken; inject glucose intraperitoneally.

6. Determine the amount of glucose in the samples by means of the Folin-Malmros method, as described in Part II, Exper. 15.

RESULTS OF TYPICAL EXPERIMENT

>The normal blood sugars, as determined by this method, run from 115 to 140 mg.; after insulin, the figures will be markedly lower, from 55 to 75 mg.

STUDENT'S REPORT

EXPERIMENT 17

ALLOXAN DIABETES

Since the pancreas in the rat is a diffuse structure, the production of diabetes by pancreatectomy is difficult. In this experiment, diabetes is produced by the injection of alloxan, a drug which has a specific destructive action on the islets of Langerhans.

MATERIALS AND EQUIPMENT

Alloxan solution containing 100 mg/cc.
Reagents for the Folin-Malmros sugar-determination method.

PROCEDURE

1 The rats used should be starved for 48 hours.

2 Anesthetize lightly with Nembutal.

3 Obtain a sample of tail blood for the normal blood sugar determination.

4 Inject alloxan intraperitoneally in a dose of 300 mg/kg.

5 After 24 hours obtain another sample of tail blood. It will be found that a marked degree of hyperglycemia has been produced. If desired, the test may be repeated the following day; the blood sugar will be found to be even higher.

RESULTS OF TYPICAL EXPERIMENT

The hyperglycemia produced by alloxan is followed by a pronounced hypoglycemia, and the rat may die within a few days in hypoglycemic shock. Glucose administration will overcome the shock; the animals can then be kept alive indefinitely with insulin.

STUDENT'S REPORT

EXPERIMENT 18

THE PHENOLSULFONEPHTHALEIN TEST FOR KIDNEY FUNCTION

It has been shown that PSP (phenol red) is removed from the circulation almost entirely by the kidney and that any damage to the kidney will decrease the rate at which the dye is excreted. This experiment is designed to demonstrate the use of PSP in the determination of kidney function.

MATERIALS AND EQUIPMENT

Instruments for cannulation.
Trachea, carotid, and femoral cannulae.
Phenolsulfonephthalein (PSP) solutions.

    Concentrations of 2 mg/cc and 0.006 mg/cc in saline with enough NaOH added to give a clear solution (3 drops of 1 N NaOH in 50 cc.).

0.1 N NaOH.
Stop watch.
Colorimeter.
Centrifuge.

PROCEDURE

1 Cannulate the trachea, left carotid artery, and right femoral vein. Heparinize.

2 Dissect out the renal artery and renal vein to one kidney and place one loop of thread around the two vessels. Do not tie.

3 Remove 0.5 cc. of blood from the carotid artery; transfer to a centrifuge tube. This sample is used to make the standard for colorimetric analyses.

4 Inject 5 mg/kg of PSP via the femoral vein; start the stop watch.

5 Exactly 2 minutes after the injection of the PSP, collect 0.5 cc. of blood and transfer to a second centrifuge tube.

6 After about 30 minutes, prevent the blood flow to one kidney by tightening the loop of thread previously placed around the renal vessels.

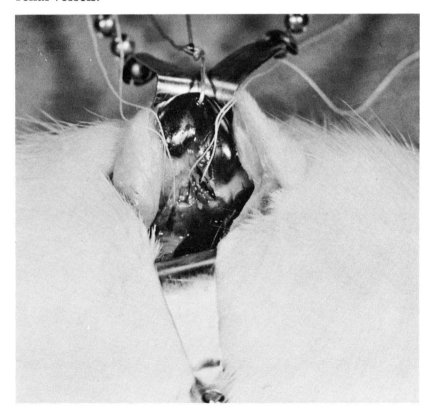

18 continued

7 Repeat the injection of 5 mg/kg of PSP, and exactly 2 minutes after the injection remove another 0.5-cc. sample of blood; transfer to a third centrifuge tube.

8 Centrifuge the three blood samples, measure 0.2 cc. of plasma from each and transfer to numbered test tubes.

9 To the first tube (standard) add 1.5 cc. of water, 1.0 cc. of the PSP solution containing 0.006 mg/cc and 3 cc. of 0.1 N NaOH. To tubes 2 and 3 add 2.5 cc. of water and 3 cc. of 0.1 N NaOH.

10 Compare the resulting solutions in a colorimeter.

11 Calculate the percentage retention of the dye when there is normal kidney function and when the circulation to one kidney has been impaired. Assume the plasma volume to be 3 per cent of the body weight.

STUDENT'S REPORT

CALCULATIONS

Weight of rat: 225 gm.

Calculated plasma volume: $225 \times 0.03 = 6.75$ cc.

$Rs = 20$ mm.

Rx (first sample) $= 28$ mm.

Rx (second sample) $= 13$ mm.

$Cs = 0.006$ mg. PSP.

$$\frac{20}{28} \times 0.006 \times \frac{6.75}{0.2} = 0.145 \text{ mg. PSP remaining after 2 minutes (normal animal).}$$

$$\frac{0.145}{1.13} \times 100 = 12.8 \text{ per cent retention.}$$

$$\frac{20}{13} \times 0.006 \times \frac{6.75}{0.2} = 0.31 \text{ mg. PSP remaining after 2 minutes (impaired kidney function).}$$

$$\frac{0.31}{1.13} \times 100 = 27.6 \text{ per cent retention.}$$

EXPERIMENT 19

THE ASCHHEIM-ZONDEK PREGNANCY TEST

This test is based on the excretion, during pregnancy, of chorionic anterior pituitary-like (A.P.L.) hormone. Its presence is shown by the reaction of the ovaries of immature (21–24-day-old) rats.

MATERIALS AND EQUIPMENT

Urine from a pregnant woman.

If no such urine is available, the instructor may prepare a substitute by dissolving the requisite amount of a commercial preparation of the chorionic hormone, such as Korotrin (Winthrop Chemical Co.), in ordinary urine. It should have a strength of approximately 100 I.U. per cubic centimeter.

Specimen of ordinary (nonpregnant) urine.

PROCEDURE

1. Both urine preparations are extracted with ether and the ether discarded (some urine samples are toxic, the toxic substances being ether-soluble). The urine samples are then filtered and kept in the refrigerator.

2. Each student is assigned four immature female rats, two are injected subcutaneously with each of the urine samples, the student not knowing which is the pregnant sample. The injection schedule is 2 injections daily of 0.5 cc. per injection for 3 days.

3. The animals are sacrificed at the end of 100 hours after the first injection, and the reproductive organs of the two groups are contrasted.

RESULTS OF TYPICAL EXPERIMENT

The chorionic hormone in the pregnant sample will have produced opening of the vaginal orifice, cornification of the vaginal epithelium, a great increase in the size of the uterus, and maturation (follicle and corpus luteum formation) in the ovaries.

The experiment may be extended by demonstrating the effect of chorionic hormone on the seminal vesicles of immature males by injecting them with the pregnant sample, according to the same injection schedule. It will be found that the seminal vesicles have increased markedly in size as compared to normal males of the same age.

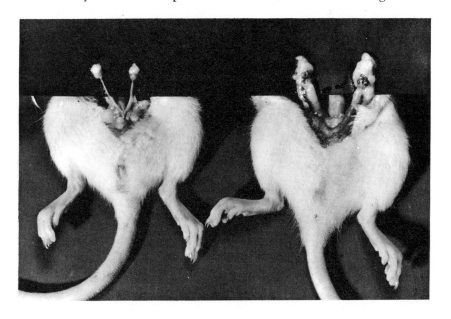

19 continued

STUDENT'S REPORT

EXPERIMENT 20

THE EFFECTS OF LONG-CONTINUED ESTRIN ADMINISTRATION

In this experiment the effect of estrin, acting presumably through the pituitary, on the reproductive system, mammary glands, and pituitary of male rats is demonstrated. Although the mechanism of action is not well understood, the results are interesting, and the experiment is included as illustrative of problems which the student may choose to attack by independent study.

MATERIALS AND EQUIPMENT

A commercial estrin preparation, as noted in Part I, Experiment 26.

PROCEDURE

1 Adult male rats, about 6 months of age, are assigned; a few are kept as controls, the rest are injected daily (they need not be injected on Sundays) with 300 I.U. of estrin. The injections are made subcutaneously; the volume should be kept small, about 0.2 cc., and the site of injection varied. Injections should be continued for a period of from 6 to 8 weeks, whereupon the animals, along with the controls, are sacrificed.

2 Estrin administration over long periods produces several remarkable effects, which are now noted. The testes and seminal vesicles are dissected out and weighed; it will be found that they have decreased greatly in size as compared with the controls. (Microscopic examination of stained sections of the testes will demonstrate almost complete degeneration of the seminiferous tubules.) The failure of interstitial tissue is shown by the small size of the seminal vesicles.

3 Male rats have only very rudimentary mammary glands and no nipples. Following this course of estrin injections, it will be found, by cutting through the skin at the site of the mammary glands, that they have become well developed, comparable to those of a lactating female, and are actively secreting a milk-like fluid.

4 The pituitaries of control and experimental animals are dissected out and weighed. To do this, the skull is removed with bone-cutters, the brain cut through just behind the olfactory lobes and lifted upward and backward. In doing this, the optic nerves and chiasma are seen and the nerves ruptured. Beneath the chiasma the pituitary is seen, inclosed in the dura. It is freed with fine-pointed forceps and removed. The effect of estrin administration is greatly to increase the size of the pituitary, often to double or three times its normal weight.

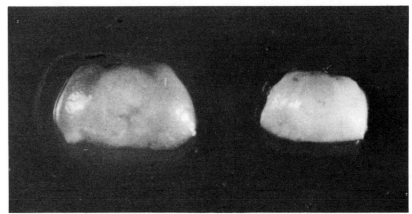

Pituitary from estrin-injected rat (*left*), and normal pituitary.

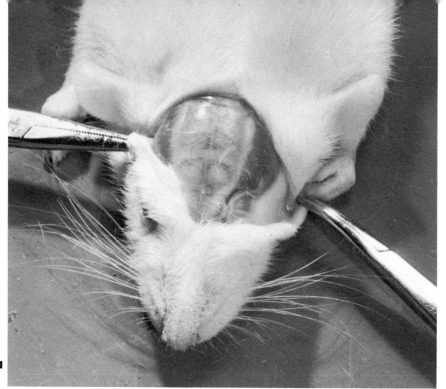
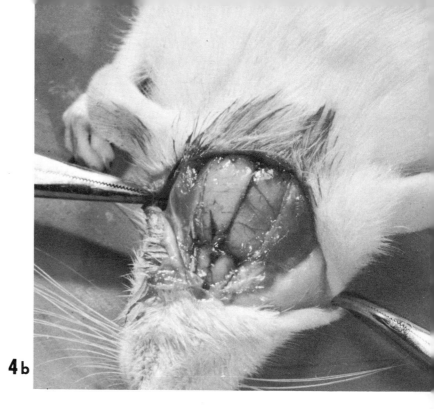
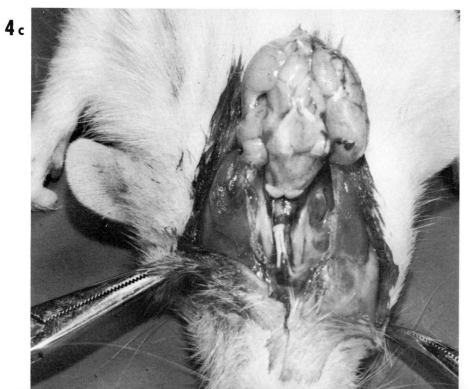
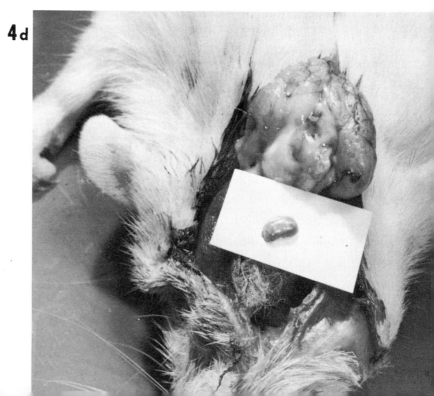

RESULTS OF TYPICAL EXPERIMENT

	Testes (gm.)	Average weight Seminal vesicles (gm.)	Pituitaries (mg.)
Controls	2.6	1.3	9.0
Injected	0.75	0.15	22.0

STUDENT'S REPORT

EXPERIMENT 2

EFFECTS OF HYPOPHYSECTOMY ON WEIGHT GAIN AND THE ESTRUS CYCLE

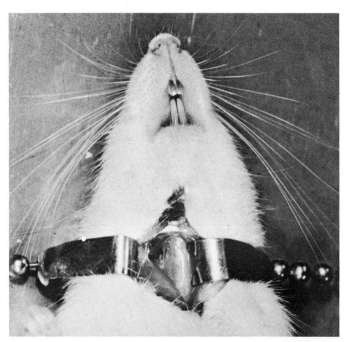

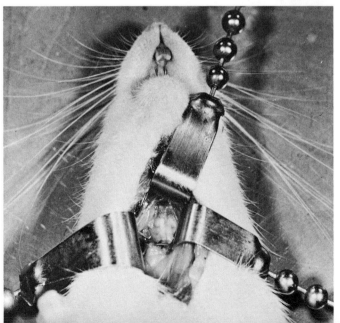

The purpose of this experiment is to show that there is a failure of normal growth and a cessation of the estrus cycle when the hypophysis is removed.

MATERIALS AND EQUIPMENT

Drill.

A drill may be improvised, using any low-speed motor, a flexible shaft, and a dental burr. The motor listed by the Central Scientific Company, Cat. No. 18802, is satisfactory. The burr should have a diameter of about 1 mm.

Surgeon's lamp.

Small bone chisel.

Suction apparatus.

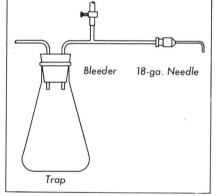

Suction apparatus for hypophysectomy

This consists of either a vacuum pump or water aspirator, a trap, bleeder, and 18-gauge blunt hypodermic needle bent at a right angle.

PROCEDURE

1. Anesthetize a 150-gm. female rat with ether (not Nembutal). Make a mid-line incision, as for tracheal cannulation. Dissect between the sternohyoid and omohyoid, at about the level of the thyroid, directing the dissection medially. Continue the dissection until the forceps touch the floor of the sphenoid. With a small bone chisel clear away the overlying muscle until the ridge, which lies medially on the sphenoid, is clear-

ly visible. Follow the ridge headward to the point where it broadens and ends. It is at this point that the hole is to be drilled.

2 Remove retractors and, holding the incision open by means of a pair of forceps, begin drilling, using low speed until the hole has been started. Avoid applying too great a pressure as the hole deepens. Usually the drill ruptures the dura, which lies between the hypophysis and the sphenoid; if not, the membrane must be opened with a fine probe. Now, using strong suction, insert the tip of the bent needle into the hole. In favorable cases the hypophysis will come out in one piece, held to the tip of the needle by suction; if not, it will be necessary to rotate the tip of the needle in all directions in order to remove all parts of the gland.

3 Close the skin incision with wound clips; no suturing of the muscle is necessary.

4 Weigh and make vaginal smears for a period of 1 month, following the weight gains of a group of controls over the same period.

RESULTS OF TYPICAL EXPERIMENT

Weight. Controls gained 30 gm. over a period of 1 month (average of 6 rats). Hypophysectomized lost 25 gm. over this period (average of 6 rats).

Estrus cycle. At the end of 2 weeks the estrus cycle had ceased in the operated animals.

STUDENT'S REPORT

EXPERIMENT 22

INFLUENCE OF THE SYMPATHETICS ON THE BLOOD VESSELS OF THE EAR

This experiment repeats the classic work of Claude Bernard, who, in 1851, demonstrated the existence of vasoconstrictor nerves.

MATERIALS AND EQUIPMENT

Hooded flashlight
 (pencil flashlight with small tube of white cardboard around end).
Inductorium.
Usual surgical instruments.

PROCEDURE

1 Expose the carotid sheath on the right side. Carefully free the cervical sympathetic nerve, which lies close to the wall of the carotid artery, running parallel to it and the vagus nerve. It is much smaller than the vagus and can be readily identified by the presence of a ganglion at the headward end, at about the level of the carotid bifurcation. Place a tie around it as far bodyward as possible and cut bodyward from the tie.

2 Examine the two ears with the hooded flashlight, the table light being turned off meanwhile. It is best to rub a little glycerin on the ears and view them by transmitted light. The ear on the side of the cut nerve will be redder, and close examination will disclose many small open vessels, especially along the central artery, not visible in the other ear.

3 Now free the nerve as far headward as possible, holding it by the thread. Lay the animal on its side so that the ear can be directly observed while the nerve is being stimulated. Stimulate the nerve, straddling the electrodes across it and using a moderately strong current. The small open vessels will close off and the ear become blanched. Cease stimulating and observe the rapid return of the blood flow.

4 Now shift the position of the animal so that the observer has a good view of the pupil of the eye. The hooded flashlight is useful for this purpose, as the illumination should not be too bright. Again stimulate the nerve and observe the dilatation of the pupil and its return to normal when the stimulus is discontinued.

STUDENT'S REPORT

EXPERIMENT 23

A STUDY OF THE SPINAL REFLEXES

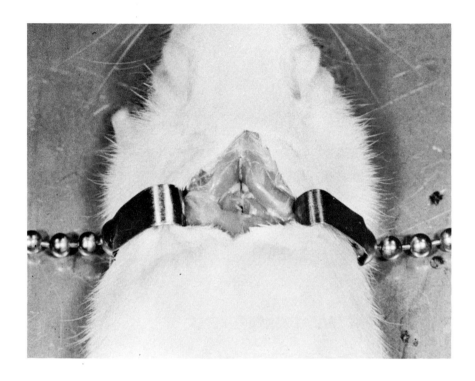

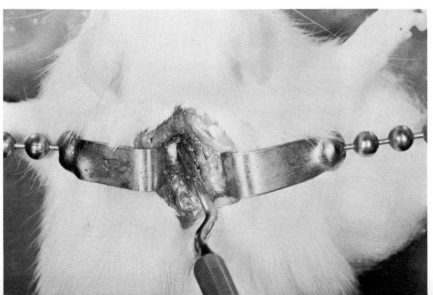

The purpose of this experiment is to determine the time of return of reflexes in the spinal rat.

MATERIALS AND EQUIPMENT

Sterile pack.
Bone-cutters.

PROCEDURE

Young, 3–5-week-old female rats are used. Females are preferable to males, as the latter develop bladder disturbances more readily.

1. Anesthetize with ether.

2. Place the animal in the prone position with a rolled towel under its abdomen. This facilitates the operative procedure and prevents interference with respiration and circulation caused by pressure against the body.

3. Make a dorsal mid-line incision in the mid-thoracic region. Clear away the muscle, applying hot saline packs to stop the bleeding. Carefully remove the laminae of 2 or 3 vertebrae, and expose the spinal cord anywhere between T_6 and T_9. Section above T_5 interferes with respiration; and section below T_{13} causes degeneration of the caudal stump and failure to regain bladder control.

23 continued

4 Cut the cord with a sharp scalpel. Completeness of section may be assured by applying tension on the tail at the time of sectioning, as the cut ends will separate by 1–3 mm.

5 Fasten the muscle with silk-thread sutures and the skin incision with wound clamps. Place the animal in a warm cage.

As bladder control is not regained for from 8 to 12 days, urine should be forcibly expelled 3 times daily by pressing upon the lower abdomen with an inward and downward motion. If bladder emptying is inadequate, edema and gangrene will develop. No difficulty will be experienced with defecation.

The animals should be tested daily for the following reflexes:

1 Ipsilateral flexion (withdrawal of foot from pinching).

2 Withdrawal of tail from pinching.

3 Withdrawal of foot and tail from cold or hot water (compare with response to water of body temperature).

4 Crossed defensive response (pinching of toe causes ipsilateral flexion and contralateral "pushing-away" movement, directed toward the stimulated foot).

5 Crossed extensor reflex (rare).

6 "Stepping" response (pinching of tail causes bilateral alternate rhythmic hip and leg movements, which eventually outlast the stimulus).

7 "Walking" movements of hind legs (alternate flexion of hips after legs have been dragged for some distance).

8 Stretch reflex.

9 Scratch reflex.

10 Defensive "shrinking" position (lifting animal by chest causes extension of legs and flexion of toes, followed by flexion of toes, hip, knee, and adducted thigh, with legs closely pressed to abdomen).

11 Mass reflex, produced by pinching of toes (rare).

RESULTS OF TYPICAL EXPERIMENT

1. Ipsilateral flexion: within 1 day.
2. Withdrawal of tail from pinching: within 1 day.
3. Withdrawal from cold or hot water: within 1 day.
4. Crossed defensive response: 2–4 days.
5. Crossed extensor reflex: not observed.
6. "Stepping" response: 3–7 days.
7. "Walking" movements of hind legs: 14 days.
8. Stretch reflex: 2–4 days.
9. Scratch reflex: 3–4 days.
10. "Shrinking" position: 7–10 days.
11. Mass reflex: 2 months.

STUDENT'S REPORT

EXPERIMENT 24

THE PRODUCTION OF DECEREBRATE RIGIDITY

This experiment demonstrates the action of the antigravity muscles in maintaining the normal body posture. When the influence of suppressor areas is abolished by decerebration, the action of facilitatory regions produces an exaggeration of the normal standing position.

Before proceeding with the present experiment, it is desirable that the student study the anatomy of a brain removed from a cadaver, after it has been hardened in alcohol or formalin.

MATERIALS AND EQUIPMENT

Drill and large burr.

    The drill described in Part II, Experiment 21, is satisfactory; a somewhat larger burr may be used.

Blunt probe.

Hot saline packs.

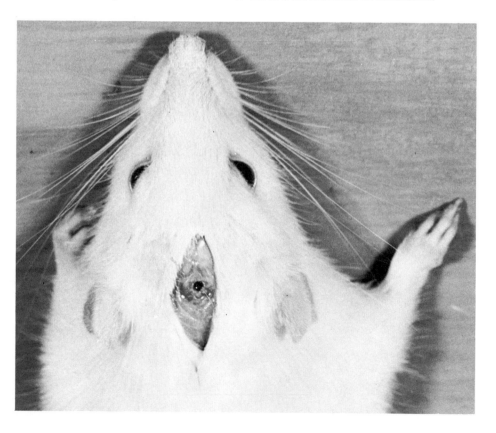

PROCEDURE

1 Anesthetize the rat lightly with ether.

2 Make a mid-line incision in the scalp over the region of the cerebellum.

3 Drill a hole through the skull in the mid-line and just caudad to the transverse sinus. (The hole will lie above the rostrad portion of the central lobe of the cerebellum.)

4 Insert the probe directly downward and, by moving its tip from side to side, sever the connections on both sides. To insure complete decerebration, move the probe back and forth several times. Transection at this level will separate all connections rostrad to the pons, leaving the vestibular nuclei, (Dieter's) intact.

5 Warm saline packs should be frequently applied to the wound until the animal appears to have recovered from the trauma. Occasionally, breathing may cease, and artificial respiration will be necessary. Rigidity begins to develop soon after recovery from the ether and reaches its maximum within about 2 hours.

STUDENT'S REPORT

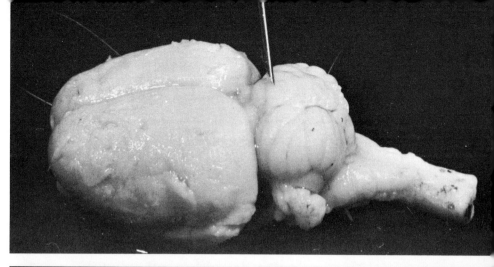

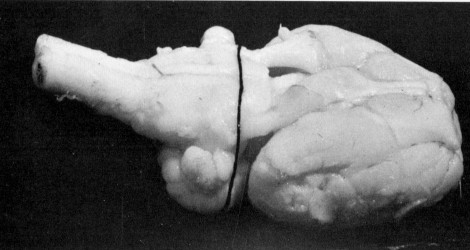

EXPERIMENT 25

EFFECTS OF LABYRINTHECTOMY

This experiment demonstrates the disturbances of equilibrium and the forced movements which result from injury to the labyrinth.

MATERIALS AND EQUIPMENT

Small bone chisel (the blade of a lancet works very well).

PROCEDURE

1 Lay the rat on its side and anesthetize with ether. Make a skin incision beginning immediately behind the ear, following the angle of the jaw.

2 Separate the masseter from the posterior digastric muscle. Pass a double ligature around the posterior facial vein which overlies these muscles, tie, and cut between the ties. The separation is deepened, and the bulla can now be felt with the points of the forceps.

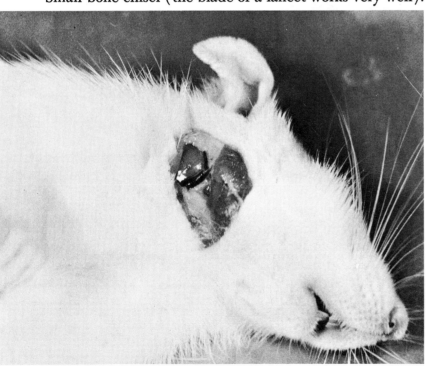

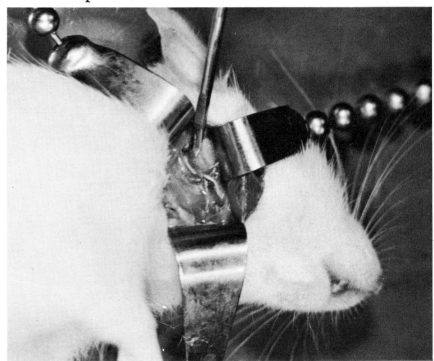

RESULTS OF TYPICAL EXPERIMENT

Unilateral labyrinthectomy

1. The eye on the operated side deviates upward, and the opposite eye deviates downward (this is the first observable result of the operation and the most certain evidence that labyrinthectomy is complete).

2. The head turns toward the operated side.

3. The animal rolls toward the operated side.

4. The legs may flex on the operated side and extend on the other side.

Bilateral labyrinthectomy

1. Position of eyes returns to normal.

2. The animal may roll toward the side of most damage. If he is kept overnight, rolling will usually have ceased by the next day.

3. Demonstrate the righting reflex in a normal rat by holding the animal in the supine position and dropping him approximately 2 feet into a box of sawdust. He will right himself and land on his feet. Repeat this procedure with the bilaterally labyrinthectomized rat. It will be found that the righting reflex is abolished.

3 Clear the bulla of overlying tissue and make a hole in it with the chisel. Then break off pieces of the bulla to enlarge the opening. The protuberance over the cochlea can now be seen. The protuberance is punctured with the chisel, and this destroys the labyrinth.

4 Close the wound with wound clips. Allow the rat to recover from the anesthetic, and record the effects of unilateral labyrinthectomy. Then repeat the operation on the other side and record the results.

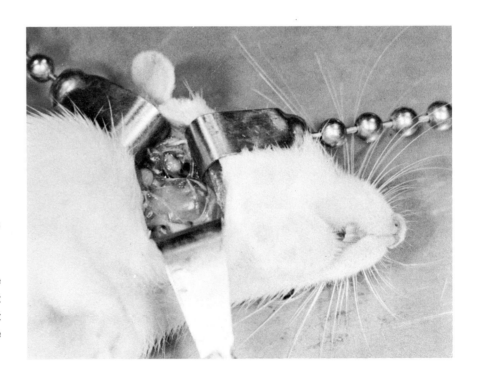

25 continued

STUDENT'S REPORT

EXPERIMENT 26

ESTABLISHMENT AND EXTINCTION OF A CONDITIONED REFLEX

In this experiment the establishment of a conditioned reflex of the "shock-avoidance" type is demonstrated; the extinction of the reflex, through use of the conditioning stimulus alone is then shown.

MATERIALS AND EQUIPMENT

Conditioning apparatus.
Inductorium and dry cell.
Buzzer.
2-way switch.

PROCEDURE

1 The rat is placed in the box for 5–10 minutes, in order to accustom himself to his surroundings.

2 Sound buzzer several times for periods of 2 seconds, to see whether the conditioning stimulus alone causes the animal to run across. (Occasionally a highly nervous rat will behave in this manner; such a rat should be discarded.)

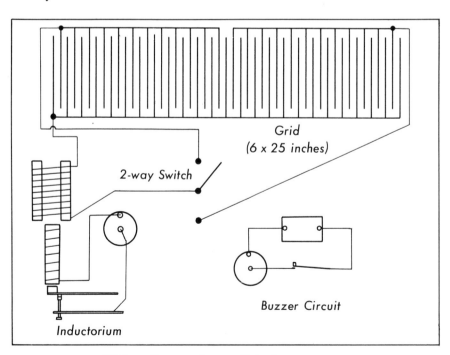

Wiring diagram for conditioning apparatus.

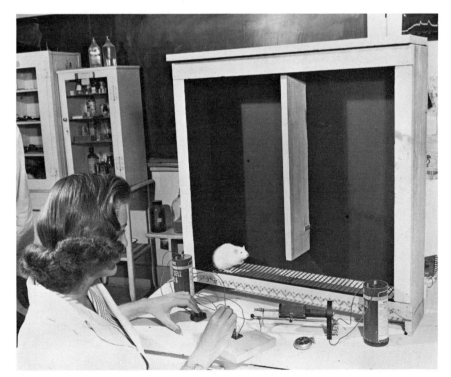

26 continued

3 Now sound buzzer for 2 seconds, and then immediately close the inductorium switch on the side of the box where the rat is resting, forcing him to run to the opposite side of the box. (The shock need not be too strong, just enough to make the rat run briskly. With a 1.5-volt dry cell, a setting of the secondary coil 4–6 cm. from its full position gives good results.) This procedure is repeated at 30-second intervals; keep a record of the rat's responses. Various criteria may be used as indicating establishment of conditioning, one being that the animal runs, on 5 consecutive trials, to the other side of the box during the 2 seconds when the buzzer is being sounded, i.e., without the shock being applied.

4 After the conditioned reflex has been established, extinction may be produced by continuing, over repeated trials, to sound the buzzer alone, until the animal fails for 5 consecutive trials to respond to it.

RESULTS OF TYPICAL EXPERIMENT

Using 5 consecutive conditioned responses as the criterion, 3 out of 5 animals required 25–30 trials, the other 2 required 40–50 trials. In all 5 rats the reflex was extinguished in 40–50 trials.

This apparatus may be used to study many aspects of conditioning and discrimination, such as the effects of endocrine imbalance, diet, removal of brain tissue, etc.

STUDENT'S REPORT

EXPERIMENT 27

ANAPHYLACTIC SHOCK

Although the rat is markedly resistant to sensitization, anaphylactic shock may almost invariably be produced by the technic described in this experiment.

MATERIALS AND EQUIPMENT

Normal horse serum.

Blood pressure setup.

PROCEDURE

1 To sensitize, inject the rats intraperitoneally with 2 cc. of horse serum per rat per injection, at 3-day intervals, for 3 injections. Fifteen days following the last injection the test for anaphylactic shock is carried out.

2 Cannulate a carotid artery and the opposite jugular vein.

3 Attach the carotid cannula to the tambour, and, while a normal record is being taken, inject 0.25 cc. of horse serum (warmed to body temperature) into the jugular vein. An extreme fall in blood pressure will occur, either immediately or within a few minutes; the heart may stop and respiration cease. The animal can sometimes be revived by administering artificial respiration.

4 To prove that sensitization was responsible, carry out a control experiment using an unsensitized rat. It will be found that the injection of 0.25 cc. of horse serum produces no such effect.

RESULTS—SAMPLE RECORDS

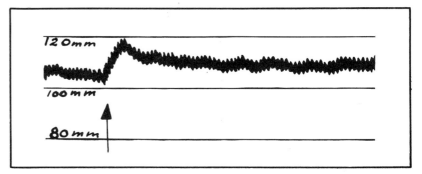

Effect of injection of 0.25 cc. of normal horse serum in unsensitized rat. The slight rise is due to increased blood volume.

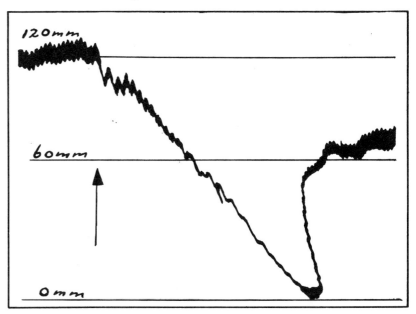

Effect of injection of 0.25 cc. of normal horse serum in sensitized rat, showing extreme fall in blood pressure.

27 continued

STUDENT'S REPORT

INTRODUCTION TO EXPERIMENTS 28 AND 29

RADIATION PROTECTION PROCEDURES AND LABORATORY REGULATIONS

The following rules are designed to conform to the regulations set forth in the Federal Register, Volume 22, Number 19, Title 10, 1957.

No unnecessary materials are to be brought into the laboratories.

Eating, drinking, and smoking in the radioisotope laboratory are forbidden.

Pipetting or the performance of any similar operation by mouth is prohibited.

Students with cuts or open wounds may not work with radioisotopes.

All radioisotope work must be localized and should be performed in vessels placed on paper or in trays.

All work involving large amounts of activity (>25 mc.) or gaseous compounds must be done in the hood.

There can be no departure from the written procedure without the permission of the instructor.

Film badges should be worn at all times while working with radioisotopes except soft beta emitters (i.e., C^{14}, S^{35}, H^3, Ca^{45}).

Pocket dosimeters should be worn at all times while working with gamma emitters. Records will be kept of all exposures.

Radioactive material should be stored so that the general background in the vicinity of the stored material is less than 1 mr/hr. All storage rooms containing radioactive isotopes and all areas in which amounts of radioisotopes in excess of counting levels are handled must be marked plainly with appropriately labeled signs.

Work should be planned so as to minimize exposure to external irradiation. In no case should the hands receive more than 100 mr/day nor the whole body more than 100 mr/week.

Rubber gloves are to be worn when working with radioisotopes at levels greater than 1 microcurie and at any time when there is danger of contact with skin.

At the end of the exercise or operation the gloves are to be washed and monitored with an appropriate monitoring instrument. Disposable gloves are to be placed in a trash can made available for radioactive wastes. The hands should then be washed and also monitored. All working areas should be cleaned and monitored for activity. Activity above background must be reported immediately to the instructor for instructions concerning further cleaning and decontamination.

Laboratory coats must be monitored thoroughly before the student leaves the laboratory.

Introduction to Experiments 28 and 29

All contaminated trash (including paper, cloth, rubber, etc.) is to be placed in the trash can painted with a red band. All contaminated glassware is to be kept in receptacles designated for these. Waste disposal and decontamination of glassware will be demonstrated by the instructor. The instructor must be consulted about waste disposal in every experiment.

No radioactive compounds or materials other than those mounted and ready for counting are to be taken into the counting room. These should be well covered while being carried from one place to another. Laboratory coats worn in the radioactive laboratories are to be removed and left outside the counting room.

Extreme care must be taken with the counters. Do not attempt to connect or disconnect any parts of the counting systems without supervision by the instructor. Do not use any instrument without previous instructions. Do not disconnect or connect any components of instruments unless the master switch is *OFF*.

When the meaning of a regulation is not clear, or a hazard not covered arises, report to the instructor for further information.

Each student is expected to do his share of washing and decontaminating equipment and of keeping working areas clean.

EXPERIMENT 28

UTILIZATION OF C¹⁴ LABELED GLUCOSE IN THE RAT

Glucose is oxidized in the animal organism to carbon dioxide and water. When D-Glucose-C^{14} is injected intraperitoneally, the carbon dioxide resulting from the glucose will be labeled and can be determined in the expired air. The rate at which oxidation of glucose occurs is dependent upon the nutritional state of the animal. A fasting animal may be expected to deposit more glucose as glycogen than a well-fed rat and thus oxidize less for energy.

MATERIALS AND EQUIPMENT

Rat chamber constructed from screw top jar.

Hardware cloth (¼-inch mesh) is fastened to lid to support rat. Commercial cage available from Atomic Accessories.

Gas washing bottles each containing 150 ml. of 20 per cent NaOH. Prepare five.

Barium hydroxide trap.

Use 5 per cent barium hydroxide ($Ba(OH)_2 \cdot 8H_2O$). 20 per cent sodium hydroxide solution—carbonate free (Fisher So-S-256).

Labeled D-Glucose-C^{14}.

Prepare solution containing 3 mg/ml in H_2O. (New England Nuclear Corp., 575 Albany Street, Boston 18, Mass., Catalog No. NEC-42L; 10^4 disintegration/second/mg.)

Barium chloride solution (5 per cent ammonium chloride, 10 per cent barium chloride).

Drying bottle containing anhydrous $CaCl_2$.

CO_2-free water.

Boil distilled water for 10 minutes, cool.

PROCEDURE

1 Set up apparatus as shown. The flow pattern of the incoming air is through the first sodium hydroxide bottle to the drying bottle, and into the animal chamber, then directly to the next two carbon dioxide absorbers. Gas is passed through the barium hydroxide trap to determine the efficiency of absorption of the carbon dioxide. Air is pulled through the system with a water aspirator.

2 Test the system for leaks.

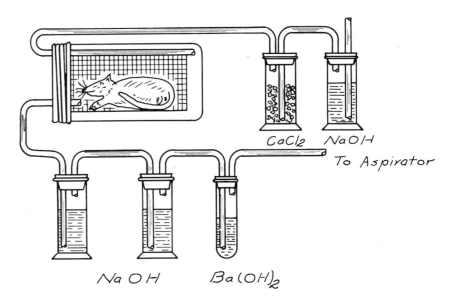

3 Inject a 100–125-gm. rat with 1 ml. of the labeled glucose. Immediately after injection place the rat in the jar and adjust the flow of air to give about 0.3–0.5 liter per minute.

4 Thirty minutes after the injection replace the two sodium hydroxide absorption bottles with two fresh bottles.

5 After an additional sixty minutes (ninety minutes after injection) remove the two sodium hydroxide bottles.

6 When the experiment has been completed, remove the rat from the chamber, place in an ether jar, and leave until dead. The rat should then be disposed of by incineration.

7 Adjust the total volume of sodium hydroxide in the two traps from the thirty-minute run and that from the sixty-minute run to 300 ml. for each.

8 To determine radioactivity, measure 10-ml. aliquots of the NaOH solution into 50-ml. centrifuge tubes. Add an excess of barium chloride–ammonium chloride solution. Centrifuge and wash precipitate with CO_2-free water. Pour off the supernatant liquid, add 2 ml. methyl alcohol and prepare a slurry of the precipitate with methanol. Transfer the methanol slurry to a weighed planchet. Allow to dry overnight at room temperature and then in an oven at 100° C. for one hour; weigh.

9 Transfer to a similar planchet an amount of ordinary glucose equal to the weight of barium carbonate found in the planchet. To this add 0.1 ml. of the labeled glucose solution used for injection. Add sufficient methyl alcohol to form a slurry of the "hot" and "cold" glucose. Dry at room temperature overnight and at 100° C. in an oven for one hour.

10 Determine radioactivity in the three planchets using a Geiger-Müller thin-window tube or a windowless gas-flow counter.

CALCULATIONS

The C^{14}-Glucose in the solution used for injection in the rat had 3×10^4 disintegrations/second/ml. Due to self-absorption during counting of the sample, all disintegrations are not measured. It is possible to correct for this by a number of technics. The procedure selected in this experiment consists of counting a known sample of the administered glucose under conditions similar to those used for the unknown.

Calculate the weight of glucose burned by the rat in the thirty-minute and sixty-minute periods.

$$\frac{\text{Counts in the barium carbonate}}{\text{Counts in the glucose planchet}} \times 3 = \text{mg. of labeled glucose oxidized.}$$

$$\frac{\text{mg. } C^{14}\text{-Glucose oxidized}}{3} \times 100 = \text{per cent } C^{14}\text{-Glucose oxidized.}$$

STUDENT'S REPORT

EXPERIMENT 29

INTESTINAL ABSORPTION AND TISSUE UPTAKE OF Co^{60}-VITAMIN B_{12}

PROCEDURE

1. Administer by stomach tube 1 ml. of Co^{60}-Vitamin B_{12} solution containing approximately 0.01 μc. and 0.05 μg. of B_{12}. This material may be purchased from the New England Nuclear Corp., 575 Albany St., Boston 18, Mass. Place the rat in a metabolism cage and collect urine and feces. Note that these animals should be kept in the special isotope room in the animal care area.

2. Collect urine and feces for three days.

3. On the third day kill the animal and wash out the gastrointestinal tract thoroughly with water and count the material separately.

4. Weigh the liver, digest a 2- to 3-gm. aliquot with 5 ml. concentrated nitric acid.

5. Digest the entire gastrointestinal tract in a similar manner.

6. A similar digest can be made of the kidneys, heart, and spleen or any other organs of interest.

7. Transfer all digests to standard size containers, dilute to 50 ml. and count in the scintillation counter.

8. Determine the activity of the urine in a similar manner.

9. Stir the fecal material in water to give a suspension and count in the same manner.

10. Assay the administered B_{12} solution by diluting an aliquot to 50 ml. and counting with the same geometry used with the samples.

11. Calculate your results in percentage of administered doses excreted in the feces and urine and percentage retained per organ studied.

IMPORTANT

1. Complete clean-up, monitoring, and decontamination procedures are the responsibility of all students as part of the experiment. See instructor for special methods.

2. A pocket ionization chamber must be worn throughout all portions of this experiment. The amount of exposure will be recorded in a notebook supplied by the instructor.

STUDENT'S REPORT

EXPERIMENT 30

TRANSAMINASE LEVELS AND TISSUE DAMAGE

Clinical experience has shown that determination of the serum levels of glutamic-oxalacetic transaminase (SGO-T) can be helpful as a diagnostic procedure when myocardial damage occurs. Transaminase enzymes catalyze the transfer of amino groups from α-amino acids to keto acids. In this experiment measurement is made of the transaminase enzyme which is active in the following reaction:

α-ketoglutartic acid + L-aspartic acid → glutamic acid + oxalacetic acid.

MATERIALS AND EQUIPMENT

Surgical instruments.

Reagents for serum glutamic-oxalacetic transaminase, SGO-T (Kit 505; Sigma Chemical Corporation, St. Louis, Mo.).

Brass rod 6 inches long, about $\frac{3}{16}$ inch in diameter, rounded on end.

Dry ice.

Spectrophotometer or colorimeter with light filter in range of 500 to 520 millimicrons.

PROCEDURE

1. Anesthetize the rat with three-quarters of a normal dose of Nembutal.

2. Collect a 1-ml. sample of blood by heart puncture. Transfer to a centrifuge tube carefully to prevent hemolysis; centrifuge the clotted blood and remove the serum.

3. Make a small opening into the thoracic cavity so as to expose the heart. Damage the heart by placing the "dry ice–cooled" rod on it.

4. Suture inter-layers of skin over the thoracic cavity to close the thorax.

5. Close the outer skin with wound clips.

6. At the end of twenty-four hours collect the second sample by heart puncture; centrifuge the clotted blood and remove serum.

7. Determine SGO-T in the two serum samples using the "Sigma" procedure.

DETERMINATION OF SGO-T (SIGMA)

1. Transfer exactly 0.2 ml. of serum to 1 ml. of Sigma prepared substrate (505–1) which has been brought to 37° C. in a water bath.

2. Exactly sixty minutes after adding the serum, add 1 ml. Sigma color reagent (505–2). Mix and allow to remain at room temperature.

3. Twenty minutes later add 10 ml. of 0.4 N NaOH to develop color. Mix by inversion.

4 Exactly thirty minutes after adding the NaOH, read in a spectrophotometer or colorimeter at a wave length of 505 (500–520) millimicrons.

5 A calibration curve must be run for the spectrophotometer. Preparation of this curve is described below.

PREPARATION OF CALIBRATION CURVE

Use calibration vials Stock No. 505–7.

1 Add 10 ml. of water to the preweighed calibration vial. Replace cap and shake to dissolve the solid material.

2 Pipette solutions into test tubes as indicated below:

Tube No.	505–7 Std. Soln. (ml.)	Sigma Prepared Substrate 505–1 (ml.)	Water (ml.)	Units SGO-T (per ml. serum)
1	0.0	1.0	0.2	0
2	.1	0.9	.2	20
3	.2	0.8	.2	55
4	.3	0.7	.2	95
5	.4	0.6	.2	148
6	0.5	0.5	0.2	216

Above values are based upon 0.2 ml. of serum.

3 Add 1 ml. of Sigma color reagent (505–2 to each tube). Shake gently and allow to stand for twenty minutes at room temperature.

4 Add 10 ml. of 0.4 N NaOH to each tube. Mix by inversion, using clean rubber stoppers.

5 Thirty minutes after adding the sodium hydroxide, read and record optical densities of all samples using water as a reference and the instrument set at 505 millimicrons.

6 Plot the calibration curve for optical density versus the corresponding units of SGO-transaminase at 37° C. This may not be a straight line.

DETERMINATION WITH COLORIMETER

If a colorimeter is used, read against standard which most nearly matches unknown.

CALCULATIONS

1. Spectrophotometer:
   Read directly from calibration curve.

2. Colorimeter:
   Calculate ratio of unknown to standard and multiply by number of units in standard sample used.

$$\frac{\text{Reading for unknown}}{\text{Reading for standard}} \times \text{conc. of standard sol.} = \text{units in unknown.}$$

RESULTS OF A TYPICAL EXPERIMENT

Normal for rat—about 100 units.

After damage to heart—150 to 200 units.

30 continued

STUDENT'S REPORT

EXPERIMENT 31

BIOSYNTHESIS OF UREA

Urea is synthesized in the animal body by the following reactions:

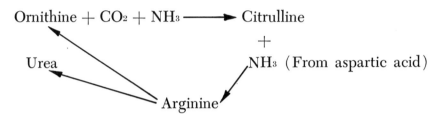

The enzymes catalyzing citrulline synthesis, arginine synthesis, and arginine hydrolysis are all present in an "acetone powder" of beef liver. This is the dry residue left after extracting liver with acetone. In this experiment the conversion of citrulline to urea will be demonstrated.

MATERIALS AND EQUIPMENT

Acetone powder, beef liver (General Biochemicals, Chagrin Falls, Ohio).

Di-potassium-L-aspartate (General Biochemicals), 210 mg/50 ml.

"Cocktail" mixture.

Dissolve the following materials in the buffer solution and dilute to 50 ml. with water:

a) 50 mg. adenosine triphosphate (Sigma Chemical Company, St. Louis, Missouri).
b) 17 mg. $MgCl_2 \cdot 6H_2O$ (Fisher Scientific).
c) 15 mg. citrulline (Sigma Chemical Company).
d) 33 ml. Tris buffer, 0.1 M, pH 7.4 (Sigma Chemical Company).

Phosphate buffer, pH 7.5, 0.1 M (Sigma Chemical Company).

5 per cent perchloric acid—CAUTION (Fisher Scientific).

Sulfuric-phosphoric acid mixture (Fisher Scientific), 25 ml. H_2SO_4 plus 75 ml. H_3PO_4.

Urea standard (Fisher Scientific), 18 mg/l.

4 per cent phenylpropanedione-2-oxime in ethyl alcohol (Fisher Scientific).

Colorimeter or photometer to read at wavelength of 550 millimicrons.

PROCEDURE

1 Extract 1 gm. of "acetone powder" with 2 ml. of 0.1 molar phosphate buffer (pH 7.5) plus 8 ml. distilled H_2O. Stir or shake the powder with the phosphate buffer to give complete extraction. Allow the suspension to stand until the solid material settles. Centrifuge if necessary.

2 Transfer the supernatant liquid to a test tube marked "enzyme solution." Place in an ice bath to preserve enzyme activity.

3 Measure 1.5 ml. of the "cocktail" mixture into each of three test tubes.

4 To test tube 2 add 0.25 ml. of water.

5 To test tubes 1 and 3 add 0.25 ml. of the potassium aspartate solution. Mix all three tubes.

6 Add 0.25 ml. of the ice cold enzyme to each tube. Mix.

7 Add 2.0 ml. of 5 per cent perchloric acid to tube 1. Mix. Tube 1 will serve as the "zero time" control.

8 Incubate the three tubes in a water bath at 37° C. for thirty minutes. (Large beaker of water is satisfactory.)

9 Stop the reaction in tubes 2 and 3 by adding 2 ml. of 5 per cent perchloric acid.

10 Shake the three tubes; centrifuge each for five minutes.

11 Decant the supernatant liquid from each into clean test tubes, each bearing the same number as the original tube.

DETERMINATION OF UREA

1 Transfer 1.25 ml. of sulfuric-phosphoric acid mixture to each of 6 Pyrex tubes.

2 Pipette 1 ml. of the *appropriate* supernatant liquid into tubes 1, 2, and 3.

3 Add 0.0, 1.0, and 2.0 ml. of the urea standard to tubes 4, 5, and 6.

4 Add distilled water to make a final volume of 4.0 ml.

5 Add 0.1 ml. of 4 per cent alcoholic solution of phenylpropanedione-2-oxime to each tube. Mix by vigorous swirling and place all tubes in a boiling water bath, darkened with India ink, for sixty minutes. Read in a colorimeter at a wavelength of 550 millimicrons. Tube 4 will serve as the reference blank.

CALCULATIONS

From the results calculate the amount of urea formed:
1. when the enzyme had no significant time to act (tube 1).
2. when the potassium aspartate was omitted (tube 2).
3. in the presence of potassium aspartate and all factors for enzyme activity (tube 3).

STUDENT'S REPORT

GLOSSARY

Aliquot. A representative part or portion of a whole.

Bulldog Clamp. Also called an "artery clip." A strong spring clamp used on blood vessels or thin-walled rubber tubing.

Calibration. The graduation of a recording instrument into known units.

Cannula. A small-bore tube of metal or glass suitable for insertion into a blood vessel or other hollow organ.

Centrifuge. A machine used to separate the heavier constituents in a suspension by rapid rotation.

Colorimeter. An instrument used in chemical analysis to determine the concentration of a colored substance in an unknown solution by comparison with its concentration in a known solution.

Hematocrit. A graduated tube used to determine the relative volumes of plasma and cellular elements in the blood. Also used to express the percentage of cells, i.e., "the hematocrit is 50."

Hemocytometer. An instrument used to count the cells in blood. It consists of a counting chamber and two pipettes, one for red cells, the other for white.

Hemolysis. The rupture of the red-cell membrane, with consequent liberation of hemoglobin.

Hemometer. An instrument used to determine the hemoglobin content of blood.

Hemostat. A type of clamp used to pinch off blood vessels or to retract tissues.

Heparin. A substance used to prevent the clotting of blood.

Hypoglycemia. A condition in which the blood sugar concentration is below normal. (Hyperglycemia is the opposite condition.)

Hypotonic. A solution having a lower osmotic pressure than blood plasma. (Hypertonic, one having a higher osmotic pressure.)

Kymograph. A revolving drum, operated by either an electric or a clock motor, used in the recording of various physiologic functions.

Laking (of Blood). The rupture of red cells and liberation of hemoglobin (same as hemolysis).

Manometer. An instrument for measuring the pressure of liquids or gases.

Milligrams per Cent Solution. Milligrams per 100 cc. of solution.

Per Cent Solution. Grams per 100 cc. of solution.

Physiologic Saline Solution. A solution of sodium chloride having the same osmotic pressure as blood (0.9 gm. NaCl per 100 cc.).

Pneumograph. An instrument for recording the rate and depth of respiration. It consists of an elongated bellows connecting with a tambour.

Signal Magnet. An electromagnet used to record the time of injection, stimulation, etc., on a kymograph drum.

Sphygmomanometer. An instrument for determining the blood pressure in the human artery.

Spirometer. A device consisting of a tank inverted in water and used to measure gas volumes, specifically, the volume of expired air.

Standard Solution. A solution of known concentration.

Stilette. A delicate probe.

Stromuhr. An instrument for measuring the velocity of blood flow.

Tambour. A drum-shaped instrument with an elastic membrane stretched across one face, by means of which pressure changes are transmitted to a recording apparatus.

Timer. A synchronous motor designed to record constant time intervals on a kymograph drum.

Vernier. A graduated scale which subdivides the smallest divisions on another scale.

NOTES

# NOTES